Emerging Topics in High-Power Laser and Light–Matter Interactions

Emerging Topics in High-Power Laser and Light–Matter Interactions

Guest Editors

Zhaohong Liu
Jiawei Sun
Sensen Li

MDPI Basel • Beijing • Wuhan • Barcelona • Belgrade • Novi Sad • Cluj • Manchester

Guest Editors
Zhaohong Liu
Center for Advanced Laser Technology (CALT)
Hebei University of Technology
Tianjin
China

Jiawei Sun
Suzhou Institute of Biomedical Engineering and Technology
Chinese Academy of Sciences
Suzhou
China

Sensen Li
College of Optical and Electronic Technology
China Jiliang University
Hangzhou
China

Editorial Office
MDPI AG
Grosspeteranlage 5
4052 Basel, Switzerland

This is a reprint of the Special Issue, published open access by the journal *Photonics* (ISSN 2304-6732), freely accessible at: https://www.mdpi.com/journal/photonics/special_issues/SBN4H0Q600.

For citation purposes, cite each article independently as indicated on the article page online and as indicated below:

Lastname, A.A.; Lastname, B.B. Article Title. *Journal Name* **Year**, *Volume Number*, Page Range.

ISBN 978-3-7258-4200-1 (Hbk)
ISBN 978-3-7258-4199-8 (PDF)
https://doi.org/10.3390/books978-3-7258-4199-8

Contents

About the Editors

Zhaohong Liu

Prof. Dr. Zhaohong Liu is a Professor at the School of Electronic and Information Engineering, Hebei University of Technology, Tianjin, China, and serves as a doctoral supervisor. He earned his Ph.D. in Optical Engineering from the School of Astronautics at the Harbin Institute of Technology in 2018. From 2021 to 2022, he completed postdoctoral research at the Center for Measurement and Sensor System Technology, Technische Universitat Dresden, Germany. Dr. Liu's research focuses on high-power laser technology, laser measurement, nonlinear optics, and light–matter interactions. His specific interests include stimulated Brillouin scattering (SBS) pulse compression, laser phase conjugation, beam shaping, and their applications in precision measurement and high-energy laser systems. Dr. Liu has authored over 40 papers in leading SCI journals such as *Applied Physics Letters* and *Optics Express*. His work has been recognized with several awards, including the First Prize for Scientific and Technological Progress from the military, the Second Prize for Technological Invention from Hebei Province, the Second Prize for Invention and Innovation of the China Association of Invention, the Second Prize of Natural Science of Weifang and the Editor-in-Chief Choice Award from *Chinese Optics Letters*. He also acts as a Guest Editor for the journal *Photonics*.

Jiawei Sun

Prof. Dr. Jiawei Sun is a full Professor at the Suzhou Institute of Biomedical Engineering and Technology (SIBET), Chinese Academy of Sciences (CAS). He earned his Ph.D. with highest honors (summa cum laude) from Technische Universität Dresden, Germany, under the mentorship of Prof. Juergen Czarske, Vice President of the International Commission for Optics and recipient of the Fraunhofer Prize. Dr. Sun's research focuses on lensless fiber-optic imaging, optical manipulation, optical tomography, and the integration of deep learning techniques in photonics. He has authored over 16 peer-reviewed journal and conference papers as the first or corresponding author, with publications in esteemed journals such as *Light: Science & Applications*, *Nature Communications*, *PhotoniX*, and *Optics Letters*. Notably, his work on quantitative phase imaging through an ultra-thin lensless fiber endoscope was among the top downloaded papers in *Light: Science & Applications* in 2022. Dr. Sun actively contributes to the scientific community as a Guest Editor for the journal *Photonics*, overseeing the Special Issue "Emerging Topics in High-Power Laser and Light–Matter Interactions". He is a member of professional organizations including SPIE, OPTICA, and IEEE, and he serves as a reviewer for numerous journals in the field.

Sensen Li

Dr. Sensen Li is a researcher affiliated with the College of Optical and Electronic Technology at China Jiliang University, located in Hangzhou, China. His previous experience includes work at the Science and Technology Electro-Optical Information Security Control Laboratory in Tianjin, China. Dr. Li's research interests encompass areas such as high-power laser systems, light–matter interactions, and optical waveform shaping. He has contributed to studies on stimulated Brillouin scattering and its applications in generating high-intensity temporal waveforms.

Preface

The field of photonics is witnessing a transformative era, propelled by advancements in high-power laser technologies and the intricate dynamics of light–matter interactions. These developments are not only deepening our understanding of fundamental physics but are also catalyzing innovations across diverse applications, from industrial processing to biomedical imaging and quantum information science.

This Special Issue of *Photonics*, titled "Emerging Topics in High-Power Laser and Light–Matter Interactions", aims to provide a comprehensive platform for researchers to showcase their latest findings and to foster interdisciplinary dialogue. The scope encompasses a broad spectrum of topics, including but not limited to high-power laser technology and applications, optical neural networks and machine learning in optics, fiber-optic sensing and communication, adaptive optics and wavefront shaping, short pulse laser generation via nonlinear optics, Brillouin scattering and its applications, solid-state laser systems, spectral imaging, and quantum imaging and quantum optics.

The motivation behind curating this Special Issue stems from the rapid advancements and the need for a consolidated repository of cutting-edge research in these domains. By bringing together contributions that span experimental, theoretical, and computational studies, we aim to highlight the current trends, challenges, and future directions in high-power laser systems and light–matter interactions.

This compilation is intended for a diverse audience, including physicists, engineers, and technologists engaged in photonics research and applications. We believe that the insights presented herein will be invaluable to both seasoned researchers and newcomers seeking to understand the evolving landscape of high-power photonics.

The realization of this Special Issue would not have been possible without the dedicated efforts of our contributors, who have shared their pioneering work. We extend our heartfelt gratitude to the reviewers for their meticulous evaluations and constructive feedback, which have significantly enhanced the quality of the published articles. We also acknowledge the editorial team at *Photonics* for their unwavering support and professionalism throughout the publication process.

As Guest Editors, we are honored to present this collection and hope it serves as a valuable resource for the photonics community.

Zhaohong Liu, Jiawei Sun, and Sensen Li
Guest Editors

photonics

Editorial

High-Power Lasers and Light–Matter Interactions

Zhaohong Liu [1,*], Sensen Li [2] and Jiawei Sun [3,4,*]

1 Center for Advanced Laser Technology (CALT), Hebei University of Technology, Tianjin 300401, China
2 College of Optical and Electronic Technology, China Jiliang University, Xueyuan Road 258, Hangzhou 310018, China; sensli@163.com
3 Suzhou Institute of Biomedical Engineering and Technology, Chinese Academy of Sciences, Suzhou 215163, China
4 Competence Center for Biomedical Laser Systems (BIOLAS), TU Dresden, Helmholtzstrasse 18, 01069 Dresden, Germany
* Correspondence: lzh@hebut.edu.cn (Z.L.); sunjiawei@sibet.ac.cn (J.S.)

High-power laser systems and the study of light–matter interactions at high intensities are crucial for numerous scientific and technological fields, ranging from industrial materials processing [1–4] and telecommunications [5–8] to fundamental physics research [9–15], biomedical applications [16–23], and advanced sensing [24–31]. Continuous advancements are focused on enhancing laser performance parameters—such as average and peak power [32–36], wavelength tunability [37–40], spectral purity [41–45], temporal pulse characteristics [46–49]—leveraging these capabilities to explore and exploit the complex interactions between intense light and various media [50–52]. This Special Issue highlights emerging developments in laser source technology, pulse manipulation, theoretical modeling, novel sensing applications, and diagnostic methods within the domain of high-power lasers and light–matter interactions.

The development of robust, high-performance laser sources tailored for specific applications remains a critical research direction. Recent efforts have focused on improving power scaling, wavelength accessibility, tunability, and beam quality. For applications requiring precise wavelength control, such as spectroscopy and coherent sensing, Sun et al. [53] proposed an external-cavity semiconductor laser design providing mode-hop-free (MHF) tuning across an extensive 140 nm range (1480–1620 nm), which covers the C + L telecommunication bands. This model achieves high spectral purity (SMSR > 61.65 dB), stable output power (>11.14 dBm), and rapid tuning (up to 200 nm/s), making it suitable for high-resolution vector spectrum analysis. Liu et al. [54] developed a kilowatt-level nanosecond laser system. Employing a hybrid MOPA architecture (fiber seed, Nd:YVO_4 pre-amplifier, and Nd:YAG slab main amplifier), they achieved an average power of 1.24 kW at a 20 kHz repetition rate, with adjustable pulse parameters (1–20 kHz, 10–300 ns), demonstrating high power extraction efficiency (39.1%). The 2 μm wavelength region is important for gas sensing, medicine, and communications application; therefore, Liu et al. [55] focused on optimizing the output of GaSb-based semiconductor lasers operating at this wavelength. By designing and implementing a slanted, wedge-shaped microlens fiber coupling scheme, they improved the output beam's symmetry and uniformity, achieving stable output power around 210 mW.

Generating short, high-energy laser pulses is another key development direction. In the field of commercial lasers, slab laser technology is naturally the most conventional technical route for obtaining high-energy lasers. In the field of ultra-short laser pulse generation, the compression technique based on nonlinear optics is a hot topic, and nonlinear optical techniques are commonly employed for pulse compression. Liu et al. [56] developed a kilowatt-level nanosecond laser system, employing a hybrid MOPA architecture (fiber seed, Nd:YVO4 pre-amplifier, and Nd:YAG slab main amplifier), They achieved 1.24 kW average power with a 20 kHz repetition rate and adjustable pulse parameters (1–20 kHz, 10–300 ns). Their system has demonstrated high power extraction efficiency (39.1%) and successfully compressed 7.4 ns input pulses down to 48.3 ps, achieving a pulse energy of 5.27 mJ and an SRS stage energy efficiency of 21.84%, thus offering a practical route for high-energy picosecond pulses. Feng et al. [57] investigated pulse compression using SBS followed by passive laser-induced breakdown (LIB) in CCl_4. They have demonstrated that doping the LIB medium with silicon carbide (SiC) nanowires effectively reduces the LIB threshold and enhances the stability of the compressed output. Their method yielded compressed pulses of 254.4 ps with an energy conversion efficiency of 34.2%.

Accurate modeling and characterization are vital for understanding fundamental processes and optimizing the performance of optical systems. For diffraction-limited imaging systems, understanding the point spread function (PSF) is crucial. Song et al. [58] derived a novel analytical solution for the defocused PSF using series expansions of confluent hypergeometric functions. This model provides insights into imaging performance degradation caused by defocusing, is independent of specific system design details, and shows good agreement (<3% RMS error) with numerical FFT methods in weak to medium defocus regimes. In laser Doppler sensors employing intersecting beams, precisely knowing the interference fringe geometry is critical for accurate measurements. Zhang et al. [59] developed and experimentally validated a high-accuracy 3D model for fringe spacing distributions in line-shaped Gaussian beam intersections. Their model, derived from Gaussian beam phase expressions, accurately predicted fringe geometry (average relative difference of <0.6% compared to the experimental results), enabling error minimization in 3D shape and velocity measurements.

High-power and precisely controlled lasers enable unique sensing modalities and material interaction studies. Wang et al. [60] explored the Goos–Hänchen shift (GHS)—a subtle lateral displacement of a reflected beam—as a potential basis for illicit drug detection. By measuring GHS variations for different substances (serum, methamphetamine, and heroin) and correlating them with complex refractive indices and dielectric constants, they have demonstrated the sensitivity of the GHS to these parameters, emphasizing its potential in the development of novel optical sensors for chemical identification. Enhancing the efficiency of photovoltaic devices is a primary scientific objective. To that end, Wang et al. [60] demonstrated broadband (400–2000 nm) spectral response enhancement in amorphous silicon p-i-n photovoltaic modules by incorporating flower-like silver nanoparticles. The observed tenfold increase in peak responsivity was attributed to the near-field plasmonic effects arising from the complex nanoparticle geometry, improving optical energy utilization. Understanding the dynamics of laser–material interactions, such as shock wave generation, is important for applications like laser peening and material processing. Li et al. [61] employed a convolutional neural network (CNN) to predict the temporal evolution of shock wave velocity, induced by the combination of millisecond–nanosecond laser pulses on silicon. The CNN model achieved high predictive accuracy ($R^2 = 0.9865$) with limited experimental data, showcasing the potential of machine learning for modeling complex, dynamic laser interaction phenomena.

Probing extreme environments, such as combustion chambers or supersonic flows, requires sophisticated diagnostic techniques. Song et al. [58] provided a timely review of the advances in femtosecond coherent anti-Stokes Raman scattering (fs-CARS) for thermometry. They highlighted the advantages of fs-CARS, such as the suppression of collisional effects and non-resonant background, thus enabling accurate, time-resolved temperature measurements crucial for understanding transient chemical reaction dynamics in harsh environments.

Overall, this Special Issue highlights several key trends in high-power laser technology and light–matter interactions. There is a continuous drive towards higher-power sources [54] and broader, more agile wavelength tuning [53], often tailored for specific application domains. Simultaneously, precise control over the temporal domain, particularly towards the achievement of ultra-short pulses via novel compression schemes [56,57], remains a key focus for accessing high peak intensities and studying ultrafast dynamics. Theoretical modeling and accurate system characterization [59] are increasingly more sophisticated, providing essential tools for optimizing design and performance. Furthermore, the exploration of subtle optical phenomena, such as the GHS for sensing [60] and harnessing plasmonics for device enhancement [61], demonstrate innovative application pathways. The integration of machine learning techniques represents a promising direction for modeling and predicting complex, nonlinear interaction dynamics where analytical solutions are intractable. Advanced diagnostics like fs-CARS are indispensable for probing the extreme conditions often generated or studied using high-power lasers. These advancements are also highlighted in this Special issue.

Future research is projected to continue following these trends, potentially exploring novel gain materials and laser architectures, pushing pulse compression limits further into the femtosecond and attosecond regimes. Moreover, by developing more sophisticated multi-physics models, which are perhaps further enhanced by machine learning, and discovering new applications based on precisely controlled light–matter interactions, this field holds tremendous potential. Integrating multiple functionalities, such as combined sensing and processing capabilities, may be another future direction. Significant progress continues to be made across the field of high-power lasers and light–matter interactions. The recent developments highlighted in this review include the realization of versatile high-power and widely tunable laser sources, innovative techniques for achieving picosecond pulses through nonlinear optics, refined analytical models for optical system performance, the application of optical phenomena and nanostructures for novel sensing and device enhancement, and the use of machine learning and advanced diagnostics to understand complex interaction dynamics. These advancements collectively underscore the role of laser technology in science and industry and point towards continued innovation in the future.

Acknowledgments: We would like to thank all the contributing authors of this Special issue for their support.

Conflicts of Interest: The authors declare no conflicts of interest.

References

1. Dutta Majumdar, J.; Manna, I. Laser material processing. *Int. Mater. Rev.* **2011**, *56*, 341–388. [CrossRef]
2. Li, L. The advances and characteristics of high-power diode laser materials processing. *Opt. Lasers Eng.* **2000**, *34*, 231–253. [CrossRef]
3. Nath, A. High power lasers in material processing applications: An overview of recent developments. In *Laser-Assisted Fabrication of Materials*; Springer: Berlin/Heidelberg, Germany, 2012; pp. 69–111.
4. Bachmann, F. Industrial applications of high power diode lasers in materials processing. *Appl. Surf. Sci.* **2003**, *208*, 125–136. [CrossRef]
5. Schneider, A.; Stillhart, M.; Günter, P. High efficiency generation and detection of terahertz pulses using laser pulses at telecommunication wavelengths. *Opt. Express* **2006**, *14*, 5376–5384. [CrossRef]
6. Knox, W.H. Ultrafast technology in telecommunications. *IEEE J. Sel. Top. Quantum Electron.* **2000**, *6*, 1273–1278. [CrossRef]
7. Jauregui, C.; Limpert, J.; Tünnermann, A. High-power fibre lasers. *Nat. Photonics* **2013**, *7*, 861–867. [CrossRef]
8. Jia, S.; Lo, M.C.; Zhang, L.; Ozolins, O.; Udalcovs, A.; Kong, D.; Pang, X.; Guzman, R.; Yu, X.; Xiao, S.; et al. Integrated dual-laser photonic chip for high-purity carrier generation enabling ultrafast terahertz wireless communications. *Nat. Commun.* **2022**, *13*, 1388. [CrossRef]
9. Mourou, G.; Tajima, T.; Bulanov, S. Optics in the relativistic regime. *Rev. Mod. Phys.* **2006**, *78*, 309–371. [CrossRef]

10. Khalatpour, A.; Paulsen, A.K.; Deimert, C.; Wasilewski, Ž.R.; Hu, Q. High-power portable terahertz laser systems. *Nat. Photonics* **2021**, *15*, 16–20. [CrossRef]
11. Jackson, S.D. Towards high-power mid-infrared emission from a fibre laser. *Nat. Photonics* **2012**, *6*, 423–431. [CrossRef]
12. Hirose, K.; Liang, Y.; Kurosaka, Y.; Watanabe, A.; Sugiyama, T.; Noda, S. Watt-class high-power, high-beam-quality photonic-crystal lasers. *Nat. Photonics* **2014**, *8*, 406–411. [CrossRef]
13. Liu, P.Q.; Hoffman, A.J.; Escarra, M.D.; Franz, K.J.; Khurgin, J.B.; Dikmelik, Y.; Wang, X.; Fan, J.Y.; Gmachl, C.F. Highly power-efficient quantum cascade lasers. *Nat. Photonics* **2010**, *4*, 95–98. [CrossRef]
14. Südmeyer, T.; Marchese, S.; Hashimoto, S.; Baer, C.; Gingras, G.; Witzel, B.; Keller, U. Femtosecond laser oscillators for high-field science. *Nat. Photonics* **2008**, *2*, 599–604. [CrossRef]
15. Xu, G.; Colombelli, R.; Khanna, S.P.; Belarouci, A.; Letartre, X.; Li, L.; Linfield, E.H.; Davies, A.G.; Beere, H.E.; Ritchie, D.A. Efficient power extraction in surface-emitting semiconductor lasers using graded photonic heterostructures. *Nat. Commun.* **2012**, *3*, 952. [CrossRef]
16. Jawad, H.J.; Sultan, A.F. Review recent developments in high-power diode lasers for biomedical applications. *J. Opt.* **2024**, 1–6. [CrossRef]
17. Allen, T.J.; Beard, P.C. High power visible light emitting diodes as pulsed excitation sources for biomedical photoacoustics. *Biomed. Opt. Express* **2016**, *7*, 1260–1270. [CrossRef]
18. Kieffer, J.C.; Fourmaux, S.; Krol, A. The ultrafast high-peak power lasers in future biomedical and medical x-ray imaging. In Proceedings of the 19th International Conference and School on Quantum Electronics: Laser Physics and Applications, Sozopol, Bulgaria, 26–30 September 2016; SPIE: Bellingham, WA, USA, 2017; Volume 10226, pp. 306–315.
19. Sun, J.; Wu, J.; Wu, S.; Goswami, R.; Girardo, S.; Cao, L.; Guck, J.; Koukourakis, N.; Czarske, J.W. Quantitative phase imaging through an ultra-thin lensless fiber endoscope. *Light. Sci. Appl.* **2022**, *11*, 204. [CrossRef] [PubMed]
20. Müller, A.; Marschall, S.; Jensen, O.B.; Fricke, J.; Wenzel, H.; Sumpf, B.; Andersen, P.E. Diode laser based light sources for biomedical applications. *Laser Photonics Rev.* **2013**, *7*, 605–627. [CrossRef]
21. Sun, J.; Kuschmierz, R.; Katz, O.; Koukourakis, N.; Czarske, J.W. Lensless fiber endomicroscopy in biomedicine. *PhotoniX* **2024**, *5*, 18. [CrossRef]
22. Sanchez, M.; Gallego, D.; Lamela, H. High current short pulse driver using a high power diode laser for optoacoustic biomedical imaging techniques. *Opt. Express* **2022**, *30*, 44954–44966. [CrossRef]
23. Sun, J.; Yang, B.; Koukourakis, N.; Guck, J.; Czarske, J.W. AI-driven projection tomography with multicore fibre-optic cell rotation. *Nat. Commun.* **2024**, *15*, 147. [CrossRef] [PubMed]
24. Killinger, D.K.; Menyuk, N. Laser remote sensing of the atmosphere. *Science* **1987**, *235*, 37–45. [CrossRef]
25. Walsh, B.M.; Lee, H.R.; Barnes, N.P. Mid infrared lasers for remote sensing applications. *J. Lumin.* **2016**, *169*, 400–405. [CrossRef]
26. Lamb, R.A. A review of ultra-short pulse lasers for military remote sensing and rangefinding. *Technol. Opt. Countermeas. VI* **2009**, *7483*, 61–75.
27. Hovis, F.E.; Rhoades, M.; Burnham, R.L.; Force, J.D.; Schum, T.; Gentry, B.M.; Chen, H.; Li, S.X.; Hair, J.W.; Cook, A.L.; et al. Single-frequency lasers for remote sensing. In Proceedings of the Solid State Lasers XIII: Technology and Devices, San Jose, CA, USA, 25–29 January 2004; SPIE: Bellingham, WA, USA, 2004; Volume 5332, pp. 263–270.
28. Kobayashi, T. Techniques for laser remote sensing of the environment. *Remote Sens. Rev.* **1987**, *3*, 1–56. [CrossRef]
29. Morton, P.A.; Morton, M.J. High-power, ultra-low noise hybrid lasers for microwave photonics and optical sensing. *J. Light. Technol.* **2018**, *36*, 5048–5057. [CrossRef]
30. Lin, S.; Wang, Z.; Qi, Y.; Han, B.; Wu, H.; Rao, Y. Wideband remote-sensing based on random fiber laser. *J. Light. Technol.* **2022**, *40*, 3104–3110.[CrossRef]
31. Bäuerle, D. *Laser Processing and Chemistry*, 4th ed.; Springer: Berlin/Heidelberg, Germany, 2011. [CrossRef]
32. Omar, A.; Hoffmann, M.; Galle, G.; Sylla, F.; Saraceno, C.J. Hybrid air-bulk multi-pass cell compressor for high pulse energies with full spatio-temporal characterization. *Opt. Express* **2024**, *32*, 13235–13248. [CrossRef]
33. Omar, A.; Vogel, T.; Hoffmann, M.; Saraceno, C.J. Spectral broadening of 2-mJ femtosecond pulses in a compact air-filled convex–concave multi-pass cell. *Opt. Lett.* **2023**, *48*, 1458–1461. [CrossRef]
34. Suzuki, A.; Kassai, B.; Wang, Y.; Omar, A.; Löscher, R.; Tomilov, S.; Hoffmann, M.; Saraceno, C.J. High-peak-power 2.1 μm femtosecond holmium amplifier at 100 kHz. *Optica* **2025**, *12*, 534–537. [CrossRef]
35. Ng, G.; Li, L. The effect of laser peak power and pulse width on the hole geometry repeatability in laser percussion drilling. *Opt. Laser Technol.* **2001**, *33*, 393–402. [CrossRef]
36. Limpert, J.; Roser, F.; Schimpf, D.N.; Seise, E.; Eidam, T.; Hadrich, S.; Rothhardt, J.; Misas, C.J.; Tunnermann, A. High repetition rate gigawatt peak power fiber laser systems: Challenges, design, and experiment. *IEEE J. Sel. Top. Quantum Electron.* **2009**, *15*, 159–169. [CrossRef]

37. Ding, D.; Lv, X.; Chen, X.; Wang, F.; Zhang, J.; Che, K. Tunable high-power blue external cavity semiconductor laser. *Opt. Laser Technol.* **2017**, *94*, 1–5. [CrossRef]
38. Huang, M.C.; Zhou, Y.; Chang-Hasnain, C.J. A nanoelectromechanical tunable laser. *Nat. Photonics* **2008**, *2*, 180–184. [CrossRef]
39. Fedorova, K.A.; Cataluna, M.A.; Krestnikov, I.; Livshits, D.; Rafailov, E.U. Broadly tunable high-power InAs/GaAs quantum-dot external cavity diode lasers. *Opt. Express* **2010**, *18*, 19438–19443.[CrossRef] [PubMed]
40. McComb, T.S.; Sims, R.A.; Willis, C.C.; Kadwani, P.; Sudesh, V.; Shah, L.; Richardson, M. High-power widely tunable thulium fiber lasers. *Appl. Opt.* **2010**, *49*, 6236–6242. [CrossRef]
41. Panarella, E. Spectral purity of high-intensity laser beams. *Phys. Rev. A* **1977**, *16*, 672. [CrossRef]
42. Dong, J.; Zhang, L.; Jiang, H.; Yang, X.; Pan, W.; Cui, S.; Gu, X.; Feng, Y. High order cascaded Raman random fiber laser with high spectral purity. *Opt. Express* **2018**, *26*, 5275–5280. [CrossRef]
43. Lin, N.; Chen, Y.; Wei, X.; Yang, W.; Leng, Y. Spectral purity systems applied for laser-produced plasma extreme ultraviolet lithography sources: A review. *High Power Laser Sci. Eng.* **2023**, *11*, e64. [CrossRef]
44. Nicolodi, D.; Argence, B.; Zhang, W.; Le Targat, R.; Santarelli, G.; Le Coq, Y. Spectral purity transfer between optical wavelengths at the 10–18 level. *Nat. Photonics* **2014**, *8*, 219–223. [CrossRef]
45. Dang, L.; Huang, L.; Shi, L.; Li, F.; Yin, G.; Gao, L.; Lan, T.; Li, Y.; Jiang, L.; Zhu, T. Ultra-high spectral purity laser derived from weak external distributed perturbation. *Opto-Electron. Adv.* **2023**, *6*, 210149. [CrossRef]
46. Račiukaitis, G. Ultra-short pulse lasers for microfabrication: A review. *IEEE J. Sel. Top. Quantum Electron.* **2021**, *27*, 1100112. [CrossRef]
47. Nolte, S.; Schrempel, F.; Dausinger, F. Ultrashort pulse laser technology. *Springer Ser. Opt. Sci.* **2016**, *195*, 1.
48. Finger, J.; Kalupka, C.; Reininghaus, M. High power ultra-short pulse laser ablation of IN718 using high repetition rates. *J. Mater. Process. Technol.* **2015**, *226*, 221–227. [CrossRef]
49. Ren, J.; Cheng, W.; Li, S.; Suckewer, S. A new method for generating ultraintense and ultrashort laser pulses. *Nat. Phys.* **2007**, *3*, 732–736. [CrossRef]
50. Günter, G.; Anappara, A.A.; Hees, J.; Sell, A.; Biasiol, G.; Sorba, L.; De Liberato, S.; Ciuti, C.; Tredicucci, A.; Leitenstorfer, A.; et al. Sub-cycle switch-on of ultrastrong light–matter interaction. *Nature* **2009**, *458*, 178–181. [CrossRef]
51. Liu, G.; Chen, W.; Xiong, Z.; Wang, Y.; Zhang, S.; Xia, Z. Laser-driven broadband near-infrared light source with watt-level output. *Nat. Photonics* **2024**, *18*, 562–568. [CrossRef]
52. Gutzler, R.; Garg, M.; Ast, C.R.; Kuhnke, K.; Kern, K. Light–matter interaction at atomic scales. *Nat. Rev. Phys.* **2021**, *3*, 441–453. [CrossRef]
53. Sun, J.; Qiu, L.; Liu, L.; Sheng, L.; Cui, Y.; Huang, L.; Pan, M.; Nian, F.; Hu, J. Output Characteristics of External-Cavity Mode-Hop-Free Tunable Laser Source in C+L Band. *Photonics* **2024**, *11*, 677. [CrossRef]
54. Liu, H.; Qiu, J.; Chen, Y.; Wang, H.; Wang, T.; Liu, Y.; Song, X.; Fan, Z. 1.2 kW, 20 kHz Nanosecond Nd:YAG Slab Laser System. *Photonics* **2024**, *11*, 297. [CrossRef]
55. Liu, Z.; Wang, J.; Li, N.; Yang, Z.; Li, S.; Li, S.; Wang, W.; Bayan, H.; Cheng, W.; Zhang, Y.; et al. A 2 μm Gallium Antimonide Semiconductor Laser Based on Slanted, Wedge-Shaped Microlens Fiber Coupling. *Photonics* **2024**, *11*, 108. [CrossRef]
56. Han, X.; Liu, Z.; Li, S.; Li, S.; Yang, Z.; Su, Q.; Zhang, Y.; Bayanheshig; Xia, Y.; Wang, Y.; et al. Pulse Duration Compression by Two-Stage Stimulated Brillouin Scattering and Stimulated Raman Scattering. *Photonics* **2024**, *11*, 104. [CrossRef]
57. Feng, L.; Zhao, Y.; Zhang, W.; Sun, D. Combined Compression of Stimulated Brillouin Scattering and Laser–Induced Breakdown Enhanced with Sic Nanowire. *Photonics* **2024**, *11*, 96. [CrossRef]
58. Song, F.; Chen, Q.; Tang, X.; Xu, F. Analytical Model of Point Spread Function under Defocused Degradation in Diffraction-Limited Systems: Confluent Hypergeometric Function. *Photonics* **2024**, *11*, 455. [CrossRef]
59. Zhang, H.; Wang, J.; Wang, S. Experimental Investigations and Modeling of Interference Fringe Geometry in Line-Shaped Gaussian Beam Intersections for Laser Doppler Sensors. *Photonics* **2023**, *10*, 1132. [CrossRef]
60. Wang, Y.; Zhou, X.; Fan, X.; Huang, X.; Zhang, L.; Wang, C. Exploration of Illicit Drug Detection Based on Goos–Hänchen Shift. *Photonics* **2023**, *10*, 1270. [CrossRef]
61. Li, J.; Zhang, W.; Li, Y.; Jin, G. Prediction of Shock Wave Velocity Induced by a Combined Millisecond and Nanosecond Laser Based on Convolution Neural Network. *Photonics* **2023**, *10*, 1034. [CrossRef]

photonics

MDPI

Article

1.2 kW, 20 kHz Nanosecond Nd:YAG Slab Laser System

Hao Liu [1,2], Jisi Qiu [1], Yanzhong Chen [1], Haocheng Wang [1], Tianqi Wang [1], Yueliang Liu [1], Xiaoquan Song [3] and Zhongwei Fan [2,*]

[1] Aerospace Information Research Institute, Chinese Academy of Sciences, Beijing 100094, China; liuhao@aircas.ac.cn (H.L.)
[2] University of Chinese Academy of Sciences, Beijing 101408, China
[3] Beijing Institute of Tracking and Telecommunication Technology, Beijing 100094, China
* Correspondence: fanzhongwei@ucas.ac.cn

Abstract: In this paper, we develop a kW-level high-repetition-rate nanosecond master oscillator power amplifier (MOPA) laser system, employing a structure of fiber, Nd:YVO_4, and Nd:YAG hybrid amplification. A tunable fiber seed source is used for adjustable pulse repetition frequency and pulse width. The Nd:YVO_4 pre-amplifier, which is dual-end-pumped, achieves high gain while maintaining good beam quality, and the high-power side-pumped Nd:YAG slab main-amplifier enables efficient power amplification. The repetition rate of the output laser can be adjusted within the range of 1~20 kHz, and the pulse width can be tuned within the range of 10~300 ns. The seed output is 6 mW at a repetition frequency of 20 kHz; we achieve an average output power of 1240 W with a total power extraction efficiency of 39.1% and single-pulse energy of 62 mJ at a pulse width of 301 ns. This parameter-controllable high-power laser holds promise for applications in the laser cleaning of complex surface contaminants.

Keywords: high-power laser; diode-pumped; optical amplifiers; lasers pulsed

1. Introduction

Laser with high repetition rate ($\geq$10 kHz), high single-pulse energy ($\geq$10 mJ), and narrow pulse width have significant application value and demands in various fields, particularly within advanced manufacturing sectors such as laser cleaning, laser processing, and laser cutting [1–3]. Additionally, it possesses a unique advantage in high-repetition rate particle image velocimetry (PIV) [4–6].

High single-pulse energy laser systems at the joules to multi-joules level have been extensively reported. However, they typically operate under low repetition rate of less than 1 kHz, exhibiting complex structures and considerable physical dimensions. For example, in 2016, our group reported an average 336 W from an Nd:YAG MOPA nanosecond laser at the repetition rate of 100 Hz, with a far-field beam spot 1.71 times the diffraction limit [7]. In 2018, Zhijun Kang et al. reported a rod amplifier laser based on Nd:YAG, with a single-pulse output energy of 1 J and a repetition rate of 500 Hz [8]. On the other hand, a multitude of institutions have reported laser systems with an extremely high repetition rate but low single-pulse energies. Chunhua Wang et al. reported a 70 kHz, 90 ps laser pulse from a double-passing end-pumped Nd:YVO_4 rod amplifier laser, with the pulse energy of 143 μJ [9]. In 2016, they utilized a microchip laser seed source with a pulse width of 95 ps, which was amplified by two stages of Nd:YVO_4 end-pumped amplifiers, resulting in an output with a repetition rate of 100 kHz and a single-pulse energy of 320 μJ [10]. Fu, X. et al. demonstrated a four-stage Nd:YVO_4 amplifier MOPA laser, with a repetition rate of 500 kHz and a single-pulse energy of 240 μJ [11]. F. Saltarelli et al. demonstrated a 350 W with 940 fs pulses at the output of an Yb:YAG thin-disk oscillator mode-locked with a semiconductor saturable absorber mirror (SESAM), with a repetition rate at 8.88 MHz and 40 μJ pulse energy [12]. Yiping Zhou et al. reported a 5 kHz sub-nanosecond master oscillator power amplifier (MOPA) Nd:YVO_4 laser system with a single-pulse energy of 4.2 mJ [13]. Yongxi Gao demonstrated a 417 W, 175 kHz, 2.38 mJ innoslab chirped pulse amplification laser [14]. Additionally, fiber lasers have achieved high levels of average power output [15–17], but their single-pulse energy is limited by the issue of fiber core damage, resulting in relatively low single-pulse energy, typically in the range of a few millijoules. Furthermore, due to the small diameter of the fiber core, nonlinear effects such as stimulated Brillouin scattering (SBS), stimulated Raman scattering (SRS), and self-phase modulation (SPM) easily occur during the amplification process. These nonlinear effects pose significant limitations on the enhancement of peak power for narrow linewidth nanosecond pulse fiber lasers [18–21]. Consequently, the realization of high-repetition-rate, high-energy frequency-doubled lasers using fiber lasers becomes extremely difficult, imposing restrictions on their application in precision measurement domains.

For example, in laser-cleaning applications, when high-repetition-rate pulsed lasers interact with material surface substances, contaminants rapidly absorb laser energy, leading to a localized high temperature zone. Subsequent pulsed lasers further increase the temperature in this area. At this point, the contaminants on the surface undergo ablation and gasification, forming keyholes, which inhibit temperature diffusion. This results in an accelerated temperature rise within the keyholes, generating intense transient ablation and gasification effects and facilitating impurity removal. Clearly, a higher repetition rate and pulse energy lead to faster cleaning speeds. The pulse duration falls within the nanosecond scale, ensuring excellent cleaning effects while preventing damage to the substrate from thermal accumulation. Additionally, for composite contaminants, which typically have complex compositions, a single-laser parameter may not achieve optimal cleaning results. Consequently, it is necessary to adjust laser parameters such as repetition rate and pulse width flexibly to address the varying interaction mechanisms between different components and the laser.

To achieve a laser system with both high repetition rates and high single-pulse energy output, as well as flexible temporal control of the pulse waveform, we have developed a Nd:YAG slab hybrid amplifier based on MOPA configuration with a fiber seed source. The pulse width and repetition rate of laser system are determined by the characteristics of the seed source, which offers the advantage of tunability in both pulse width and repetition rate. This design allows for versatile applications in fields such as laser cleaning and PIV diagnostics, addressing the limitations of traditional solid-state/fiber lasers.

In this configuration, the seed source comprises a fiber laser that provides a low-power signal light in the milliwatt range. The pulse waveform is dynamically controlled through the utilization of an acousto-optic modulator (AOM). For pre-amplification, a four-stage double-end-pumped Nd:YVO_4 laser amplifier is employed, capitalizing on the advantageous characteristics of the large stimulated emission cross-section exhibited by Nd:YVO_4 crystals. This design enables a substantial amplification of the seed light, thereby enhancing the overall energy efficiency of the laser system while ensuring exceptional beam quality. In the main amplification stage, a two-stage high-power continuous-wave (CW) side-pumping scheme is implemented using a parallelogram-shaped Nd:YAG slab laser amplifier. By capitalizing on the high energy storage capability inherent in the slab amplifier, the laser system achieves a remarkable output of high-power laser amplification. In this work, 1240 W of average power is obtained, corresponding to the single-pulse energy of 62 mJ at the pulse width of 300 ns. Furthermore, the pre-compensation technique for the seed pulse waveform enabled flexible control of the laser temporal profile, making it highly promising for applications in laser cleaning.

2. Experimental Setup

The experimental setup of the 1.2 kW, 20 kHz, 300 ns MOPA laser system is shown schematically in Figure 1; the laser seed source consists of a butterfly-packaged distributed feedback (DFB) laser diode, a fiber acousto-optic modulator, and two stages of Yb-doped fiber amplifiers. This configuration provides a highly stable seed laser output with adjustable repetition rate (1~20 kHz) and pulse width (10~300 ns). The average power of the seed is 6 mW at a repetition rate of 20 kHz. The seed laser is collimated by lens L1 and then enters a beam splitting and isolation protection unit consisting of a half-wave plate, Faraday rotator (FR), and polarizing beam splitter (PBS). It is then focused by lens L2 and directed into the LD dual-end-pumped pre-amplification module PA-1. This configuration optimizes the mode matching between the signal beam and the pump beam distribution within the gain medium. The HR-coated flat mirror M1 reflects the amplified laser beam. It goes back into the PA-l for the second-passing extraction. This process achieves significant gain amplification of the weak seed signal laser while maintaining high beam quality. After that, the polarization of the main beam is modified using a half-wave plate and a Faraday rotator. This allows it to pass through the PBS and enter the beam combining unit consisting of L3 and L4. It then goes through the pre-amplification modules PA-2, PA-3, and PA-4 in sequence. After three stages of unidirectional amplification, the output power reaches around ~100 W, providing enough pre-amplification power for the subsequent main power amplification stage. After the pre-amplification stage, the beam passes through a 4F spatial filter unit composed of L5, L6, and SF1 for spatial shaping and filtering. The beam is expanded with a ratio of 1:4. It is then further expanded and collimated in the y-direction with a 1:4 beam expansion using the lens group L7 and L8 before entering the main amplification stage. The main amplification stage consists of two side-pumped Nd:YAG slab laser amplification modules, A-1 and A-2. The beam is incident at the Brewster angle and undergoes multiple total internal reflections (TIR) inside the slab, propagating in a zigzag pattern. After being reflected by angle mirrors, the beam passes through the slab crystal for the second time with different incident directions, eliminating the gain blind spots present in the zigzag path and extracting the stored energy within the amplification module more effectively. By combining uniform pumping light and efficient thermal management design, high-efficiency power extraction and amplification are achieved, resulting in kilowatt-level output power.

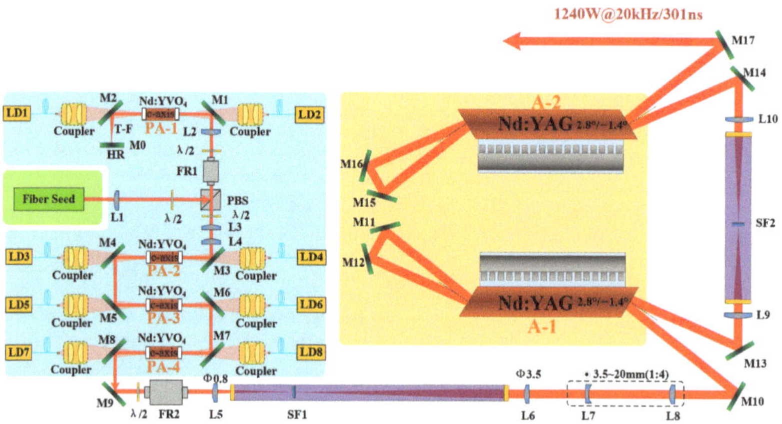

Figure 1. Experimental setup of laser system.

2.1. Pre-Amplifier Setup

The pre-amplification unit consists a four-stage end-pumped Nd:YVO_4 laser amplifier. The 0.3 at.% doped a-cut composite Nd:YVO_4 crystal (provided by Crylaser Inc., Chengdu, China) with dimensions of 4 mm × 4 mm × 20 mm is used in the amplifier as the gain medium. Nd:YVO_4 is an efficient four-level laser gain medium with a pump light centered around 808 nm and an output laser centered at 1064 nm. Compared to the Nd:YAG crystal, the Nd:YVO_4 crystal exhibits a stimulated emission cross-section approximately five times larger. This implies that the Nd:YVO_4 crystal can achieve a higher gain factor under the same operating conditions, making it an ideal material for small-signal amplification. Both end faces of the Nd:YVO_4 crystal are anti-reflection (AR), coated at 808 nm and 1064 nm, with transmittance rates of 99.9% and 98%, respectively. The crystal is encased in a 0.1 mm thick indium foil and bonded to a copper heat sink, which is cooled by circulating water at a constant temperature of 20 °C. M1~M8 are AR coated at 808 nm and HR coated at 1064 nm at an incidence angle of 45°.

Due to the relatively low output power of 6 mW from the laser seed source, it falls within the regime of a small signal for amplifiers. Consequently, even with a high gain coefficient, the energy storage of the gain medium cannot be fully utilized. In order to solve this issue, the first-stage amplifier adopts the double-pass amplification to make full use of the energy storage in the gain medium. Additionally, a coupling lens, consisting of two lenses, is employed to efficiently couple the output signal beam into the crystal. To maintain control over beam quality post-amplification, selecting an appropriate beam overlap ratio becomes crucial. This factor represents the ratio of the signal beam's diameter to that of the pump beam.

All pump sources are fiber-coupled 808 nm laser diode modules with a maximum output power of 50 W and the laser linewidth (FWHM) is 4 nm (provided by DILAS). By adjusting the temperature of the cooling water, the central wavelength can be controlled to achieve effective matching with the absorption peak of the crystal. The fiber has a core diameter of 400 μm and a numerical aperture (NA) of 0.22. The pumping light is coupled into the crystal through the two end faces of the crystal by adjusting the coupling lens group. The waist size is optimized for output parameters in each amplifier stage based on experimental results.

Figure 2 shows the schematic diagram of pump light shaping and transmission. It is appropriate to have a beam overlap ratio of around 80~90% for the signal beam and pump beam, as it allows for the effective amplification of the signal beam power while maintaining good beam quality. Therefore, the design of the lens-coupling module adopts a variable focus structure, which allows for adjusting the pump beam diameter according to different beam overlap ratio requirements.

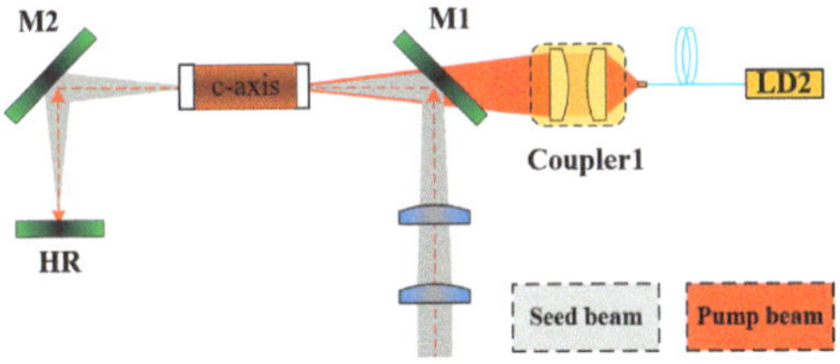

Figure 2. Illustration of the filling ratio between the seed beam and the signal beam.

According to the simulation results from Figure 3, the beam diameter of the pump light at a depth of 2 mm on the crystal end face is 400 μm for the first amplification module, with a pump power of 70 W. For the subsequent three stages, the pump light beam diameter is 600 μm, with a pump power of 90 W. A coupling lens group, consisting of two bonded lenses, is used to focus the beam into the crystal with beam-expansion ratios of 1:1 and 1:1.5, respectively. Finally, lenses L3 and L4 are used to expand and shape the amplified signal beam, ensuring that the beam passing through the last three amplifiers is nearly collimated and achieving an appropriate overlap ratio.

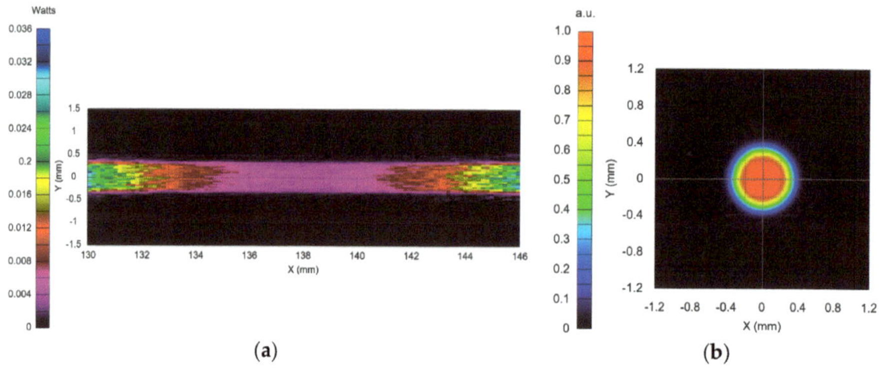

Figure 3. (**a**) Longitudinal distribution of the optical field in the Nd:YVO_4 crystal and (**b**) Transverse distribution of the beam at the focal plane of the pump light.

2.2. Main-Amplifier Setup

The main amplification unit consists of a two-stage CW side-pumped Nd:YAG slab amplifier. The 0.6 at.% slab crystal (provided by Crylaser Inc.) adopts a ~60° cut angle parallelogram structure, with dimensions of 125 mm × 22 mm × 3.5 mm (width × thickness × length). Coating the surface of the slab with a layer of SiO_2 protective film can prevent light leakage caused by the sealing ring. Currently, optical silicones of good quality have refractive indices between 1.41 and 1.53, with industrial transparent silicones generally having higher refractive indices. Assuming the lowest refractive index of 1.41, the critical angle for total internal reflection of 1064 nm light from Nd:YAG material to transparent silicone is 50.86°. When the sealing ring or sealant is in close contact with the slab, it does not disrupt the original total internal reflection condition between YAG-SiO_2. The crystal is pumped from a single large surface, and the cooling is done through the same two large surfaces of the crystal. Figure 4 shows a schematic diagram of the laser transmission, pumping region, and cooling water distribution inside the slab amplification module. The laser enters the crystal slab from the center of the end face and then undergoes total internal reflection on the upper and lower surfaces of the slab, propagating along a zigzag path. Finally, it exits from the center of the opposite end face of the slab. The large surfaces of the slab crystal are coated with an evanescent wave protective film, which increases the damage threshold. Pumping light enters the crystal slab through one of its large surfaces and is reflected by the opposite surface, effectively being absorbed completely along the thickness direction of the crystal, as indicated by the yellow region in the diagram. The blue region represents the cooling water layer. Deionized water was used as the cooling fluid, with a water layer thickness of 2 mm, and cooled at a flow rate of 30 L/min at 20 °C.

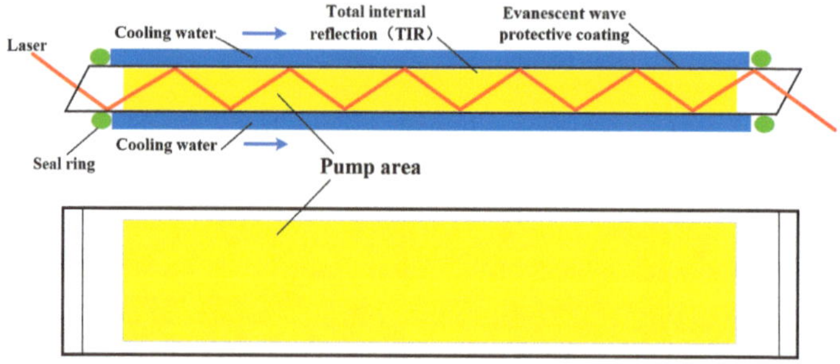

Figure 4. Schematic diagram of the laser beam path, pumping region, and cooling layer in the slab amplification module.

The slab amplifier module uses a horizontal pumping scheme, as shown in Figure 5. It consists of six vertically stacked arrays, with each array containing eight laser diode bars. Each bar has a power of 100 W (provided by Focuslight company, Xi'an, China), resulting in a maximum pumping power of 4800 W (nominal, with redundancy in reality). In the slow axis direction of the laser diode bar, the pump light is collimated by six cylindrical lens and overlaps on the crystal to form a uniformly intense optical spot. In the fast axis direction, the pump light passes through the cylindrical lens and enters a quartz waveguide. After multiple reflections on the upper and lower surfaces of the quartz waveguide, it finally combines inside the crystal to form a uniform optical spot.

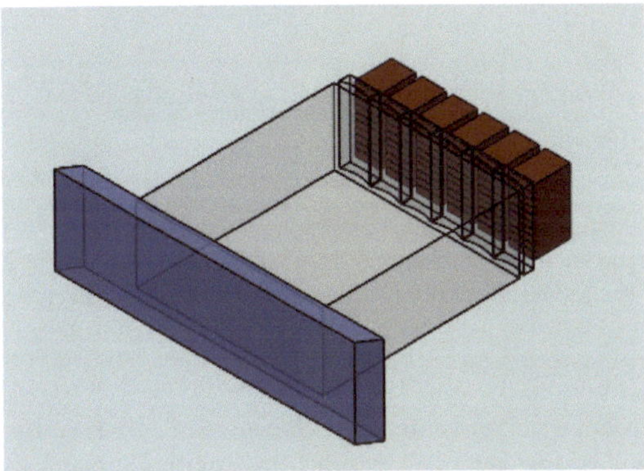

Figure 5. Schematic diagram of the side-pumping structure.

By optimizing parameters such as the spacing between laser diodes, the curvature of the cylindrical lens, and the length of the waveguide, based on simulation results, the uniformity of the pump light distribution on the crystal's cross-section exceeds 90%, as shown in Figures 6 and 7a. According to the simulation results, the spacing of the laser diode array used in our experiments is 1.7 mm; the curvature of the cylindrical lens is 15 mm, with a distance of 88 mm from the emitting surface of the laser diode; and the waveguide length is 80 mm. The dimensions of the pump region are 80 mm × 20 mm. The pump uniformity inside the slab crystal at different pump currents was tested using the image transfer method, as shown in Figure 7b.

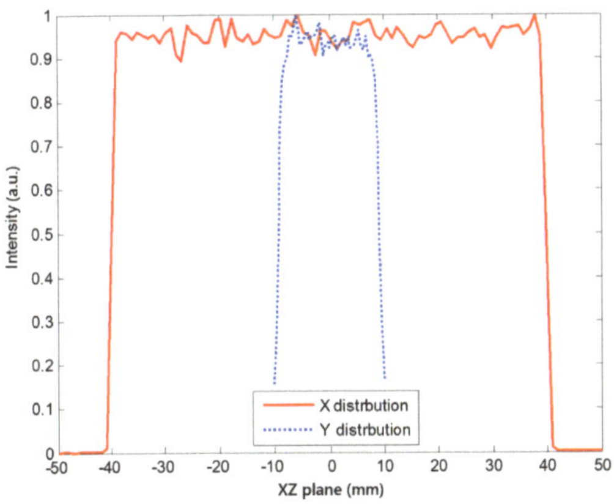

Figure 6. Transverse and longitudinal cross-sectional distribution of the pump light in the slab crystal.

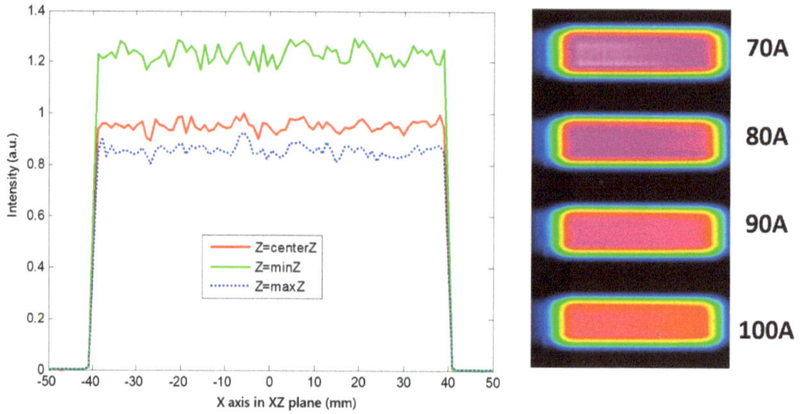

(**a**) (**b**)

Figure 7. (**a**) Transverse cross-sectional distribution of the pump light on the front surface, center, and rear surface of the slab crystal and (**b**) Test results of pump light distribution inside slab crystal at different currents.

3. Experimental Results

Figure 8a shows the average laser power and extraction efficiency of each amplifier stage. The output power of the seed source is 6 mW. By adjusting the mode matching between the signal light and the pump light of the end-pumped amplifier, the seed light is finally amplified through the first-stage double-end-pumped Nd:YVO_4 preamplifier, resulting in an output power of 10.8 W. The total gain reaches 1800 times, as shown in Figure 8b. It is worth noting that such high gain makes it prone to self-oscillation between the crystal end face and the high reflectivity (HR) mirror. By adjusting the angle of the HR mirror to be tilted by ~3° with respect to the crystal end face, the self-oscillation light can be eliminated. Then, after three stages of preamplification, the output powers are 27.6 W, 51.8 W, and 80.5 W, respectively. The power extraction efficiencies of each preamplifier stage are as follows: 28.8%, 37.3%, 64.9%, 52.7%.

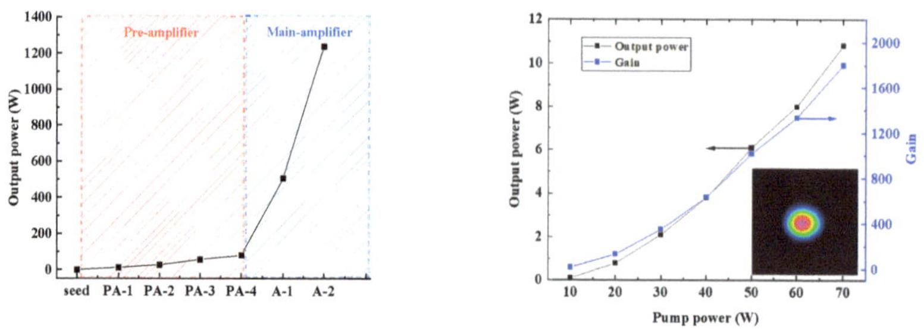

(**a**) (**b**)

Figure 8. (**a**) The output power of each amplifier stage and (**b**) output power and gain of the first-stage pump amplifier.

The energy storage, gain, and depolarization test results of the continuous pump slab amplifier are shown in Figure 9. When the injection current is 100 A, the pump power reaches 5000 W, with an output power of 1500 W, corresponding to an optical-to-optical efficiency of 30%. The single-pass gain is 1.8, with a depolarization loss of 2.75%. Due to the depolarization loss and diffraction effects in the pre-amplification stage, the laser power entering the slab amplifier is limited to 75 W. Before entering the first-stage slab amplifier, the laser beam is shaped into dimensions of approximately 3.5 mm × 20 mm. Between the two-stage slab amplifiers, there is a set of imaging transfer mirrors (L9 and L10) and a vacuum filter SF2 to suppress high-frequency components in the beam. After passing through the first and second stages of Nd:YAG slab amplifiers, the off-axis double-pass (incident angles were 2.8° and −1.4°, respectively) output powers are 506 W and 1240 W, with power extraction efficiencies of 28.4% and 48.9%, respectively. The maximum single-pulse energy is 62 mJ, with a repetition frequency of 20 kHz and a pulse width of 301 ns. The oscilloscope traces of the pulse series as well as the single pulse are illustrated in Figure 10a. The stability (RMS) of the output average power within 10 min is 0.83%. The beam quality diffraction limit of the laser output was estimated at 7 and 3, in slab width and thickness direction, respectively, due to severe thermal lensing; the far-field beam profile is shown in Figure 10b.

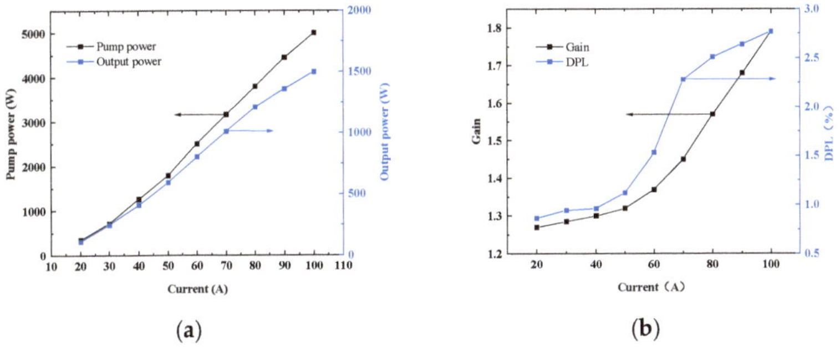

Figure 9. (**a**) Output power and (**b**) gain and bias current relationship curves for the slab amplifier module.

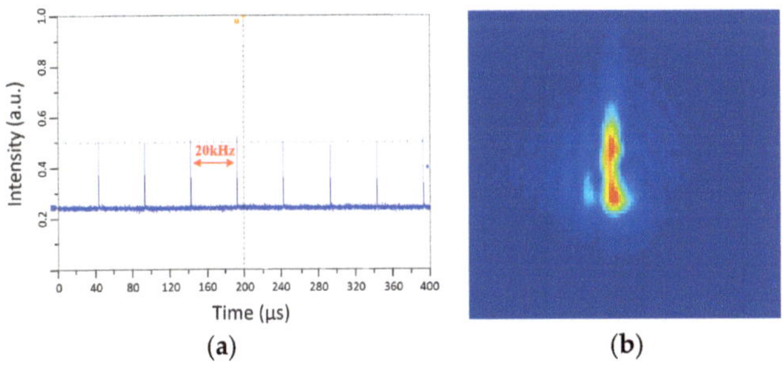

Figure 10. (**a**) The repetition frequency of the laser system output and (**b**) the far-field beam profile.

As the pulse width is 300 ns, the curves illustrating the variation of average output power and single-pulse energy at different repetition rates are depicted in Figure 11a. At a repetition rate of 1 kHz, the laser system yields an average power of 186 W, achieving a maximum single-pulse energy of 186 mJ. At 20 kHz, a maximum average output power of 1240 W is obtained. With the increase in repetition frequency, it is evident that the output power demonstrates a rising trend that gradually tends to stabilize. This can be explained by the fact that, as the pulse frequency increases, the pulse interval decreases, leading to a reduction in energy losses caused by spontaneous emission. The energy stored in the amplifier approaches a steady state. As the frequency continues to increase, pulse amplification eventually transitions into a steady-state amplification process. When the repetition rate is set to 20 kHz, the variation of average output power and single-pulse energy at different pulse widths is illustrated in Figure 11b. At a pulse width of 10 ns, the average output power is 880 W, with a single-pulse energy of 44 mJ. At 301 ns, the maximum single-pulse energy of 62 mJ is achieved.

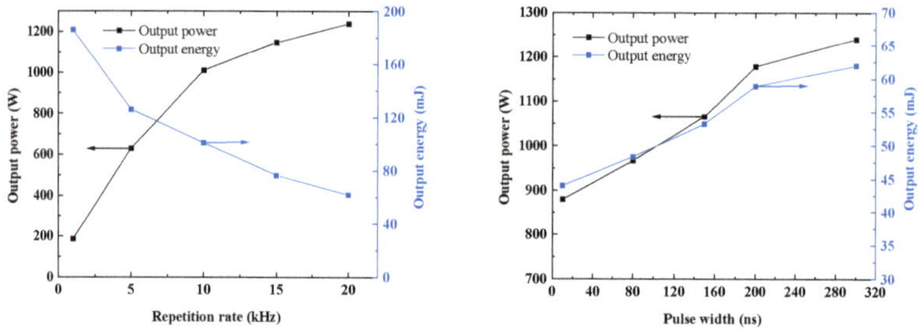

(a) (b)

Figure 11. (**a**) Output power and pulse energy versus repetition rates and (**b**) output power and pulse energy versus pulse width.

After the laser pulse waveform from the seed source is amplified by the amplifier, the presence of gain-saturation effects causes the leading edge of the pulse waveform to become relatively steep, and the pulse width is correspondingly compressed and narrowed. To address this issue, we employ the pre-shaping of the seed source output waveform to compensate for the distortion of the pulse-leading edge caused by gain saturation. As a result, a relatively smooth pulse waveform output is obtained, as shown in Figure 12, with pulse widths of 10 ns and 301 ns (FWHM), according to the test results.

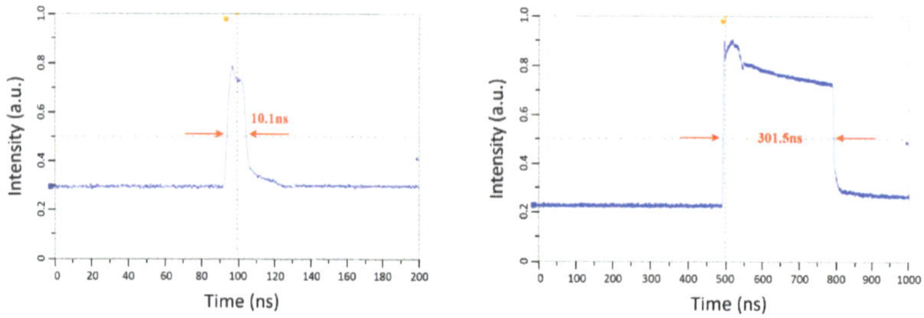

(a) (b)

Figure 12. Waveform profile of the laser system output with a typical pulse width of (**a**) 10 ns and (**b**) 300 ns.

4. Discussion

Laser systems have been developed with higher average powers (20 kW) [22], shorter pulse widths (fs) [23], longer pulse widths (QCW) [22], larger single-pulse energies [24], improved self-oscillation suppression [25], and higher beam quality [26]. However, reference [22] reports a slab-based laser oscillator with an output power level of 20 kW, but a low repetition rate of only 400 Hz and a relatively wide pulse width of 200 μs. Reference [23] describes a slab-based MOPA laser system with a high repetition rate of 20 MHz, an average power of 1.1 kW, a pulse width of 615 fs, but a corresponding single-pulse energy of only 55 μJ. In reference [24], the laser system achieves a high single-pulse energy of 800 mJ at 1064 nm, but with a low repetition rate of 400 Hz. The parasitic oscillation suppression technique described in reference [25] requires complex preparation processes like bonding crystals to achieve good results. The method described in this paper achieves parasitic oscillation suppression by tilting the crystal angle during low-power amplification, a simpler and more feasible approach. Reference [26] reports a laser oscillator based on end-pumped $Nd:YVO_4$, achieving a beam quality factor M^2 of 1.3 with an output power of only 28.8 W, which is relatively easy to achieve in laser systems with low power output. Additionally, Comaskey et al. [27] reported a laser oscillator based on an LD array side-pumped slab amplifier, achieving output powers in the kilowatt range under continuous operation with an optical–optical efficiency of 19%, while the slab amplifier we developed achieved an optical–optical efficiency of 30%.

At high repetition rates, achieving high-energy and narrow-pulse-width lasers simultaneously is a challenging research endeavor. Currently, there are two technical approaches to achieve this goal: bulk lasers based on the MOPA configuration [28,29] and fiber lasers [30]. However, the pulse waveform of conventional bulk lasers lacks flexible control, and fiber lasers operating at high repetition rates often exhibit limited single-pulse energy, typically not exceeding 10 mJ. The laser system we report achieves a single-pulse energy of 62 mJ at 20 kHz. Furthermore, the repetition rate, pulse width, and waveform are subject to flexible adjustments.

5. Conclusions

We have reported a kW level high-repetition-rate nanosecond laser, employing a structure of fiber, Nd:YVO_4, and Nd:YAG hybrid amplification. The pulse width and repetition frequency of the fiber seed source can be flexibly adjusted. By precisely matching the beam overlap ratio of the double-end-pumped Nd:YVO_4 pre-amplifier, we obtained a high-gain laser output. Finally, a dual-pass off-axis extraction scheme was used through the high-uniformity side-pumped Nd:YAG slab laser. The seed output was 6 mW at a repetition frequency of 20 kHz; we achieved an average power output of 1240 W with the total power extraction efficiency of 39.1% and a single-pulse energy of 62 mJ at the pulse width of 301 ns. Furthermore, the pre-compensation technique for the seed pulse waveform enabled flexible control of the laser temporal profile, making it highly promising for applications in laser cleaning. The final indicators are advanced in the current research field, offering the potential to bring new perspectives to the development of related areas.

Author Contributions: Conceptualization, H.L. and J.Q.; methodology, H.L., J.Q., Y.C. and X.S.; software, T.W.; validation, H.W., T.W. and Y.L.; formal analysis, H.W.; data curation, H.W.; writing—original draft preparation, H.L.; writing—review and editing, H.L.; supervision, Z.F.; project administration, Z.F.; funding acquisition, J.Q. and Z.F. All authors have read and agreed to the published version of the manuscript.

Funding: This work was supported by the Project: National Nature Science Foundation of China program (22227901).

Institutional Review Board Statement: Not applicable.

Informed Consent Statement: Not applicable.

Data Availability Statement: Data underlying the results presented in this paper are not publicly available at this time but may be obtained from the corresponding author upon reasonable request.

Conflicts of Interest: All authors declare no conflicts of interest.

References

1. Kumar, A.; Sonar, V.R.; Das, D.K.; Bhatt, R.B.; Behere, P.G.; Afzal, M.; Kumar, A.; Nilaya, J.P.; Biswas, D.J. Laser cleaning of tungsten ribbon. *Appl. Surf. Sci.* **2014**, *308*, 216–220. [CrossRef]
2. Singh, A.; Choubey, A.; Modi, M.H.; Upadhyaya, B.N.; Oak, S.M.; Lodha, G.S.; Deb, S.K. Cleaning of carbon layer from the gold films using apulsed Nd:YAG laser. *Appl. Surf. Sci.* **2013**, *283*, 612–616. [CrossRef]
3. Jang, D.; Oh, J.H.; Lee, J.-M.; Kim, D. Enhanced efficiency of laser shock cleaning process by geometrical confinement of laser-induced plasma. *J. Appl. Phys.* **2009**, *106*, 014913. [CrossRef]
4. Adrian, R.J. Twenty years of particle image velocimetry. *Exp. Fluids* **2005**, *39*, 159–169. [CrossRef]
5. Hsu, P.S.; Roy, S.; Jiang, N.; Gord, J.R. Large-aperture, tapered fiber–coupled, 10-kHz particle-image velocimetry. *Opt. Express* **2013**, *21*, 3617–3626. [CrossRef] [PubMed]
6. Fu, C.; Yang, X.; Li, Z.; Zhang, H.; Yang, Y.; Gao, Y. Experimental investigation on an acoustically forced flame with simultaneous high-speed LII and stereo PIV at 20 kHz. *Appl. Opt.* **2019**, *58*, C104–C111. [CrossRef] [PubMed]
7. Qiu, J.S.; Tang, X.X.; Fan, Z.W.; Wang, H.C.; Liu, H. Two-beam combined 3.36 J, 100 Hz diode-pumped high beam quality Nd:YAG laser system. *Appl. Opt.* **2016**, *55*, 5630–5633. [CrossRef]
8. Kang, Z.; Fan, Z.; Huang, Y.; Zhang, H.; Ge, W.; Li, M.; Yan, X.; Zhang, G. High-repetition-rate, high-pulse-energy, and high-beam-quality laser system using an ultraclean closed-type SBS-PCM. *Opt. Express* **2018**, *26*, 6560–6571. [CrossRef]

9. Wang, C.; Liu, C.; Shen, L.; Zhao, Z.; Liu, B.; Jiang, H. 1.6 MW peak power, 90 ps all-solid-state laser from an aberration self-compensated double-passing end-pumped Nd:YVO$_4$ rod amplifier. *Appl. Opt.* **2016**, *55*, 2399–2403. [CrossRef]
10. Wang, C.; Shen, L.; Zhao, Z.; Liu, B.; Jiang, H.; Chen, J.; Liu, D.; Liu, C. 95-ps all-solid-state laser with a low-power microchip laser seed and a two-stage single-pass bounce geometry amplifier. *J. Opt. Soc. Am. B* **2016**, *33*, 884–890. [CrossRef]
11. Fu, X.; Liu, Q.; Yan, X.; Cui, J.; Gong, M. 120 W high repetition rate Nd:YVO$_4$ MOPA laser with a Nd:YAG cavity-dumped seed laser. *Appl. Phys. B* **2009**, *95*, 63–67. [CrossRef]
12. Saltarelli, F.; Graumann, I.J.; Lang, L.; Bauer, D.; Phillips, C.R.; Keller, U. Power scaling of ultrafast oscillators: 350-W average-power sub-picosecond thin-disk laser. *Opt. Express* **2019**, *27*, 31465–31474. [CrossRef]
13. Zhou, Y.; Li, X.; Wei, C.; Chen, X.; Xu, H.; Fan, R.; Chen, D.; Jiang, Y.; Yan, R. 5 kHz, 4.2mJ, 900 ps end-pumped Nd:YVO$_4$ MOPA laser system. *Opt. Express* **2022**, *30*, 29833–29840. [CrossRef] [PubMed]
14. Gao, Y.; Guo, J.; Huang, Y.; Gao, Z.; Gan, Z.; Tu, Z.; Liang, X.; Li, R.; Li, R. 417 W, 2.38 mJ Innoslab amplifier compressible to a high pulse quality of 406 fs. *Opt. Lett.* **2023**, *48*, 5328–5331. [CrossRef] [PubMed]
15. Platonov, N.; Yagodkin, R.; De La Cruz, J.; Yusim, A.; Gapontsev, V. Up to 2.5-kW on non-PM fiber and 2.0-kW linear polarized on PM fiber narrow linewidth CW diffraction-limited fiber amplifiers in all-fiber format. In Proceedings of the Fiber Lasers XV: Technology and Systems, San Francisco, CA, USA, 26 February 2018.
16. Tian, X.; Rao, B.; Wang, M.; Xi, X.; Wang, C.; Li, H.; Wang, Z. 3.5 kW narrow-linewidth FBG-based MOPA fiber lasers with high signal to noise ratio. In Proceedings of the Thirteenth International Conference on Information Optics and Photonics (CIOP 2022), Xi'an, China, 15 December 2022; Volume 12478, pp. 629–632.
17. Meng, D.; Ma, P.; Wang, X.; Ma, Y.; Su, R.; Zhou, P.; Yang, L. Kilowatt-level, high brightness, narrow-linewidth PM fiber amplifiers based on laser gain competition. In Proceedings of the Fifth Symposium on Novel Optoelectronic Detection Technology and Application, Xi'an, China, 12 March 2019; Volume 11023, pp. 864–869.
18. Otto, H.-J.; Jauregui, C.; Limpert, J.; Tünnermann, A. Average power limit of fiber-laser systems with nearly diffraction-limited beam quality. In Proceedings of the Fiber Lasers XIII: Technology, Systems, and Applications, San Francisco, CA, USA, 9 March 2016; Volume 9728, pp. 82–87.
19. Distler, V.; Möller, F.; Strecker, M.; Palma-Vega, G.; Walbaum, T.; Schreiber, T.; Tünnermann, A. High power narrow-linewidth Raman amplifier and its limitation. In Proceedings of the Fiber Lasers XVII: Technology and Systems, San Francisco, CA, USA, 21 February 2020; Volume 11260, pp. 8–13.
20. Ke, W.-W.; Wang, X.-J.; Bao, X.-F.; Shu, X.-J. Thermally induced mode distortion and its limit to power scaling of fiber lasers. *Opt. Express* **2013**, *21*, 14272–14281. [CrossRef] [PubMed]
21. Sincore, A.; Bodnar, N.; Bradford, J.; Abdulfattah, A.; Shah, L.; Richardson, M.C. SBS Threshold Dependence on Pulse Duration in a 2053 nm Single-Mode Fiber Amplifier. *J. Light. Technol.* **2017**, *35*, 4000–4003. [CrossRef]
22. Guo, Y.; Peng, Q.; Bo, Y.; Chen, Z.; Li, Y.; Zhang, L.; Shao, C.; Yuan, L.; Wang, B.; Xu, J.; et al. 24.6 kW near diffraction limit quasi-continuous-wave Nd:YAG slab laser based on a stable–unstable hybrid cavity. *Opt. Lett.* **2020**, *45*, 1136–1139. [CrossRef] [PubMed]
23. Russbueldt, P.; Mans, T.; Weitenberg, J.; Hoffmann, H.D.; Poprawe, R. Compact diode-pumped 1.1 kW Yb:YAG Innoslab femtosecond amplifier. *Opt. Lett.* **2010**, *35*, 4169–4171. [CrossRef]
24. Li, S.; Ma, X.; Li, H.; Li, F.; Zhu, X.; Chen, W. Laser-diode-pumped zigzag slab Nd:YAG master oscillator power amplifier. *Chin. Opt. Lett.* **2013**, *11*, 071402.
25. Sridharan, A.K.; Saraf, S.; Sinha, S.; Byer, R.L. Zigzag slabs for solid-state laser amplifiers: Batch fabrication and parasitic oscillation suppression. *Appl. Opt.* **2006**, *45*, 3340–3351. [CrossRef]
26. Ma, Z.; Li, D.; Hu, P.; Shell, A.; Shi, P.; Haas, C.R.; Wu, N.; Du, K. Monolithic Nd:YVO$_4$ slab oscillator-amplifier. *Opt. Lett.* **2007**, *32*, 1262–1264. [CrossRef]
27. Comaskey, B.J.; Albrecht, G.F.; Beach, R.J.; Velsko, S.P.; Sutton, S.B.; Mitchell, S.C.; Petty, C.S.; Jancaitis, K.S.; Benett, W.J.; Freitas, B.L.; et al. One-kilowatt average-power diode-pumped Nd:YAG folded zigzag slab laser. In Proceedings of the Diode Pumping of Average-Power Solid State Lasers, Los Angeles, CA, USA, 16 June 1993.
28. Negel, J.-P.; Loescher, A.; Bauer, D.; Sutter, D.; Killi, A.; Ahmed, M.A.; Graf, T. Second generation thin-disk multipass amplifier delivering picosecond pulses with 2 kW of average output power. In *Advanced Solid State Lasers*; Optical Society of America: San Diego, CA, USA, 2016; p. ATu4A-5.
29. Dietz, T.; Jenne, M.; Bauer, D.; Scharun, M.; Sutter, D.; Killi, A. Ultrafast thin-disk multi-pass amplifier system providing 1.9 kW of average output power and pulse energies in the 10 mJ range at 1 ps of pulse duration for glass-cleaving applications. *Opt. Express* **2020**, *28*, 11415–11423. [CrossRef]
30. Zhu, C.; Hu, I.-N.; Ma, X.; Galvanauskas, A. *Single Mode 9.1 mJ and 10 ns Pulses from 55 μm Core Yb-Doped CCC Fiber MOPA*; CLEO: San Jose, CA, USA, 2013; pp. 1–2.

photonics

MDPI

Communication

Output Characteristics of External-Cavity Mode-Hop-Free Tunable Laser Source in C+L Band

Jisheng Sun [1,2], Liqiang Qiu [3], Lei Liu [2], Liwen Sheng [2,4,5,*], Yudong Cui [6], Lin Huang [2,4,5], Mengchun Pan [1], Fushun Nian [2,4,5,*] and Jiafei Hu [1]

1 College of Intelligence Science and Technology, National University of Defense Technology, Changsha 410073, China
2 Ceyear Technologies Co., Ltd., Qingdao 266555, China; linhuang@zju.edu.cn (L.H.)
3 Beijing Institute of Control and Electronic Technology, Beijing 100000, China
4 Science and Technology on Electronic Test & Measurement Laboratory, Qingdao 266555, China
5 Shandong Electronic Test & Measurement Technology Innovation Center, Qingdao 266555, China
6 State Key Laboratory of Modern Optical Instrumentation, College of Optical Science and Engineering, Zhejiang University, Hangzhou 310027, China; cuiyd@zju.edu.cn
* Correspondence: shengliwen2008@163.com (L.S.); nianfushun@ceyear.com (F.N.)

Abstract: Tunable laser sources with a wide wavelength tuning range, mode-hop-free (MHF) operation, and high spectral purity are essential for applications such as high-resolution spectroscopy, coherent detection, and intelligent fiber sensing. In this paper, we present a wide-range tunable laser source that operates without mode hopping, based on external cavity feedback using a semiconductor gain chip as the laser gain medium. The wavelength, power, and spectral characteristics of the laser are experimentally measured. A wide MHF continuous wavelength tuning range from 1480 nm to 1620 nm with a side-mode suppression ratio of more than 61.65 dB is achieved. An output optical power of more than 11.14 dBm with good power stability can also be realized in the full C+L band. This proposed external-cavity tunable laser source features a narrow intrinsic linewidth and MHF tunable radiation with a maximum sweep speed of 200 nm/s, enabling practical applications such as high-resolution vector spectrum analysis.

Keywords: external-cavity; tunable laser source; mode-hop-free; C+L band

1. Introduction

The development of the first laser [1], using a ruby crystal as the gain medium, marked a milestone in the academic community, sparking extensive research into new laser technologies [2–9], particularly semiconductor diode lasers that utilize a semiconductor as the gain medium [10–12]. Semiconductor diode lasers (also known as diode lasers) have opened avenues for numerous applications, including frequency and timing metrology [13–15], laser cooling [16], coherent optical communication [17], and optical fiber sensing [18–21]. Additionally, it is worth mentioning that semiconductor lasers now serve as efficient pump sources for crystal gain media [22–24]. Across these fields, external-cavity tunable laser sources (EC-TLSs) with flexible and controllable emission characteristics have become the primary consideration in the above application fields. The demands of the ever-growing mode-hop-free (MHF) tuning range have resulted in the development of various novel external-cavity configurations. Up to now, the structure of EC-TLSs has been classified into the two most common types, namely the Littrow and Littman–Metcalf geometries. An EC-TLS with a wide tuning range of 173 nm based on the Littrow geometry was reported by K. Fedorova et al., where the output spectral linewidth and side mode suppression ratio (SMSR) are less than 400 pm and 50 dB at around 1550 nm, respectively [25]. Wang et al. demonstrated that the diode laser chip in conjunction with an external optical feedback provided by a high first-order diffraction efficiency (about 91%) dispersion element can significantly improve the characteristics of the EC-TLS based on the Littrow structure and obtain a 209.9 nm tunable range [26]. Although wide tunable ranges have been obtained, little attention has been paid to the MHF output characteristics in the work mentioned above. In addition, the 0th-order lasing output beam steers used as the dispersion element (usually grating) are rotated in order to change the lasing wavelength. Therefore, the application range of the Littrow cavity laser is limited.

Compared with the classical Littrow cavity, the Littman–Metcalf configuration incorporates an additional mirror. Tuning in the Littman–Metcalf configuration is achieved by mechanically rotating the mirror angle rather than adjusting the incidence angle of the dispersion unit. This design allows the output beam direction to remain fixed as the lasing wavelength is tuned. The Littman–Metcalf cavity can be designed to increase the tuning range in which continuous MHF tuning is possible. An EC-TLS based on the traditional Littman–Metcalf design was reported by Gong et al., where the maximum MHF tunable range of 78 GHz was measured operating at 774.5 nm [27]. The proposed EC-TLS operated under an unoptimized reflector pivot position. An about 59.13 nm continuous MHF tuning range EC-TLS with a Littman–Metcalf geometry based on a diffraction grating was reported by Zhang et al., with an optical signal to noise ratio (OSNR) of around 65 dB and a spectral linewidth of less than 100 kHz [12]. By corotating a periscope with an etalon and a 0.8 nm (full width half maximum) narrow bandpass filter inside a modified Littman–Metcalf configuration, Zhu et al. demonstrated a novel MHF tunable EC-TLS without mechanical pivot-point tuning with a maximum linewidth of around 1 MHz, where the proposed Littman–Metcalf cavity had a MHF continuous tuning range of 1.7 THz [28]. At present, many scholars are inclined to the study of EC-TLSs based on the Littman–Metcalf geometry with a wider tuning range, but the research of continuous wavelength tuning with MHF is rarely reported.

In this study, we present a MHF laser system with a wide continuous wavelength tuning range based on the traditional Littman–Metcalf geometry, maintaining the narrow output spectral linewidth and high SMSR advantages of typical EC-TLSs. The output characteristics of the designed EC-TLS are investigated in detail. Experimentally, a continuously tunable range of about 140 nm with no mode hopping is achieved, which is an about 71.4% times enhancement compared with the reference [9], and the SMSR over the entire wavelength tuning output range is estimated to be better than 61.65 dB. Using self-heterodyne interferometry, the laser intrinsic linewidth is measured to be less than 5.37 kHz at the emitting wavelength of 1570 nm, and the maximum output power of the designed EC-TLS is about 15.95 dBm. In addition, a high power stability of ± 0.024 dB/1 h and wavelength stability of ± 3.08 pm/1 h are obtained, respectively. The significant advantage of this proposed EC-TLS is its ability to tune the lasing wavelength finely over a wide range without mode hopping, thus making it suitable for practical applications such as ultra-high spectral analysis systems.

2. Configuration of the EC-TLS and Device Characteristics

The configuration of the proposed laser is illustrated in Figure 1. The commercially available gain chip (Thorlabs, SAF1150S2) employs a multi-quantum well and ridge waveguide structure, achieving an output bandwidth of 133.3 nm centered at a typical wavelength of 1521.7 nm. The length of the gain chip is about 1 mm, and it is coated with a 0.05% anti-reflection film on its angled facet and a 10% reflectivity film on the other side. This gain chip is powered by a custom-built, ultra-low-noise current driver capable of supplying a maximum output current of 400 mA. Additionally, it is cooled to approximately 25.0 °C using a 7.56 W thermoelectric cooler. The output laser beam from the gain chip is shaped by a molded aspherical lens (Edmund, #87-155). The plano side and aspherical side of the collimating lens are coated with anti-reflection film, its numerical aperture is 0.55, and the focal length is 4.51 mm. The shaped laser beam is incident on the surface of the optical feedback element. Feedback is provided by a gold-coated blazed diffraction grating with 900 lines per mm (Newport Richardson, 33025FL01-155R). The average diffraction efficiency into the 1st-order is around 75% if the polarization of the gain chip is oriented perpendicular to the lines of the grating. The blazed diffraction grating, measuring 15 mm × 6 mm × 6 mm, provides a sufficient tuning range for diffracted wavelengths without clipping the lasing edges of the optical feedback element. The first-order diffracted beam is retroreflected by an added mirror back into the active layer, where it is amplified. The tuning of the selected mode is achieved by adjusting the added mirror, which is driven by a brushless direct current (DC) motor, relative to the incident beam. The fiber in this setup transmits the tunable laser signals and connects with other test instruments through port A to characterize the output characteristics of the proposed EC-TLS. As we have reported in ref. [29], to expand the wavelength tuning range of the proposed EC-TLS, here we adopt a new collimation method that requires the distance between the waist position and molded aspherical lens to be equal to the length of the external cavity.

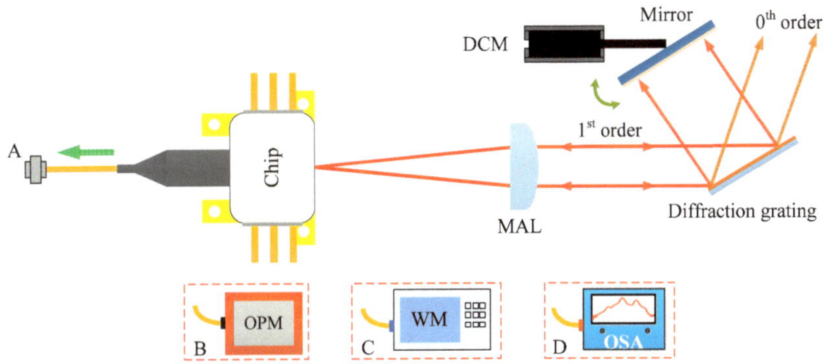

Figure 1. Schematic drawing of the proposed EC-TLS. MAL, molded aspherical lens; DCM, direct current motor; OPM, optical power meter; WM, wavelength meter; OSA, optical spectrum analyzer.

The B port of the optical power meter (Ceyear, 6337D) is connected to output port A to characterize power-related properties such as the threshold current, peak output power, and power stability. By connecting the C port of the wavelength meter (Yokogawa, AQ6151B) or the D port of the optical spectrum analyzer (Ceyear, 6362D) with a set resolution of 10 pm, wavelength-related characteristics including the tuning range and wavelength stability can be measured.

3. Results and Discussion

After constructing the experimental system, the performance of the laser source with the added external cavity was initially observed. The typical P-I curves of the laser source, optimized by external optical feedback, are shown in Figure 2. Figure 2a depicts the P-I plot when the resonant output wavelength is in the short wavelength region. The threshold current is approximately 180 mA at 1480 nm, and the maximum output power reaches 13.01 mW (11.14 dBm). Figure 2b shows the P-I curve at 1550 nm, indicating a threshold current of about 70 mA and a maximum output power of approximately 39.39 mW (15.95 dBm), a significant increase compared to the gain chip's output power of 0.12 mW without external optical feedback (as per the supplier's test report). The experimental results demonstrate that the laser system operates under strong optical feedback conditions and achieves optimal alignment. Figure 2c shows that at 1620 nm, the threshold current is about 190 mA, and the maximum output power is approximately 22.13 mW (13.45 dBm). The observed variation in the threshold current, decreasing and then increasing as the lasing wavelength changes from 1480 nm to 1620 nm, is primarily due to the spectral gain profile of the gain chip. To achieve a wider tuning range and higher output power, the operating current is set to 400 mA, leveraging the drive capability of the homemade laser diode driver, which is well above the maximum threshold current.

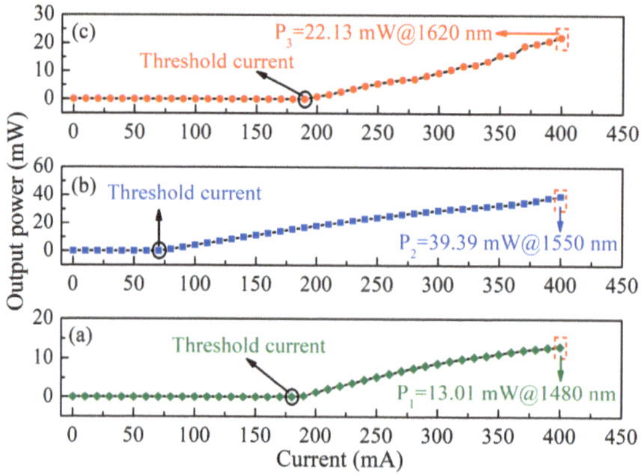

Figure 2. P-I curves of the laser source: (**a**) P-I plot of the laser source operating at 1480 nm; (**b**) P-I plot of the laser source operating at 1550 nm; (**c**) P-I curve of the laser source operating at 1620 nm.

Meanwhile, the relationship between the lasing wavelength and the injection current is studied. As shown in Figure 3, the maximum slope is only 0.313 pm/mV, which implies that the proposed laser source has strong robustness, that is, mode hopping occurs when the injection current fluctuates by more than 80 mV instantaneously. In fact, this phenomenon is different from the experimental results we observed before, and we infer that the probability of this result is due to the difference in the parameters of the gain chip itself.

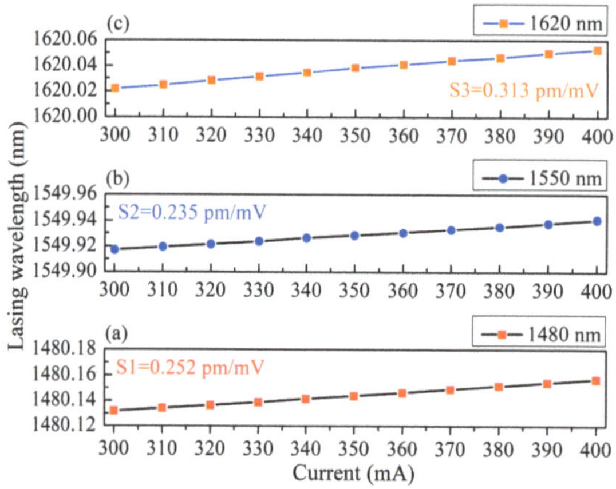

Figure 3. Lasing wavelength as a function of the injection current: (**a**) wavelength–injection plot of the laser source operating at 1480 nm; (**b**) wavelength–injection plot of the laser source operating at 1550 nm; (**c**) wavelength–injection curve of the laser source operating at 1620 nm.

Figure 4a shows the spectra of the tunable laser source at an injection current of 400 mA. It can be seen from the figure that at this time, the lasing output wavelength of the EC-TLS can be changed from 1480 nm to 1620 nm with an about 20 nm/step; that is, the tuning range is about 140 nm, and the spectrogram shows a single longitudinal mode operation, with a minimum SMSR of about 61.65 dB. The output optical power at different lasing wavelengths is then tested using a self-developed power meter produced by Ceyear. As seen in Figure 4b, the output optical power varies with the lasing wavelength. The maximum and minimum output powers of the EC-TLS are 15.95 dBm and 11.14 dBm, respectively. Clearly, with an injection current of 400 mA, an output power exceeding 11.14 dBm can be achieved across the entire C+L band.

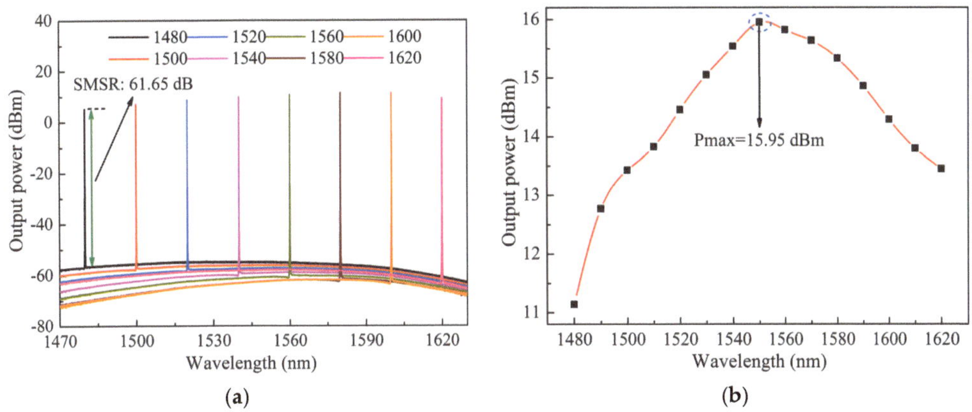

Figure 4. (**a**) Spectrogram of the proposed EC-TLS at different resonant output wavelengths; (**b**) the relationship between the output power and the resonant output wavelength.

For our proposed EC-TLS, the maximum axial mode spacing is close to 26.2 pm operating at 1620 nm, which is due to the designed cavity length of about 50 mm. In practice, the smaller the axial mode spacing, i.e., the closer to 21.9 pm, the shorter the resonant output wavelength. Therefore, when the first-order or second-order difference in the adjacent output wavelength test data is less than the minimum axial mode spacing, it indicates that the laser system operates in a MHF tunable output throughout the entire band. As displayed in Figure 5a, according to the test results, the maximum and minimum wavelength differences are calculated to be 16.12 pm and −1.6 pm, respectively. Meanwhile, an experiment is carried out to investigate the sweep speed based on the classical optical frequency domain reflectometer (OFDR) scheme [21]. Figure 5b shows that the tuning range of 80 nm is 0.401 s, that is, the sweep speed is estimated at 200 nm/s. In addition, the inset plotted in Figure 4b indicates that the proposed EC-TLS can operate in a MHF state because no phase mutations are observed.

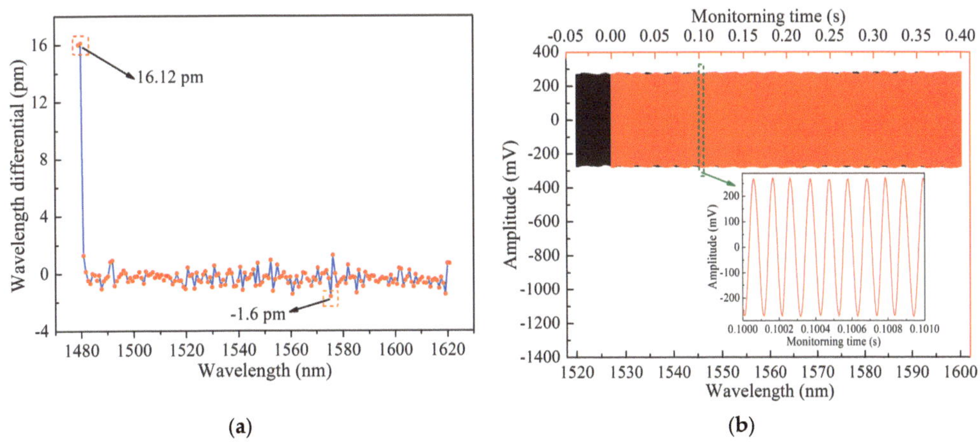

Figure 5. (**a**) MHF performance of the proposed EC-TLS; (**b**) sweep speed performance of the proposed tunable laser source.

The corresponding output wavelength and power stability curves of the tunable laser source in the case of one hour of continuous output (sample rate: 1 Hz) at an injection current of 400 mA and a controlled ambient temperature of 22 ± 1 °C are shown in

Figures 6a and 6b, respectively. According to the results, the peak-to-valley or root-mean-square (RMS) values of the wavelength stability and output power of the laser configuration can be calculated as ±3.08 pm/1 h or ±0.961 pm/1 h and ±0.024 dB/1 h or ±0.0095 dB/1 h, respectively, indicating that the tunable laser source has good wavelength and output power long-term stability.

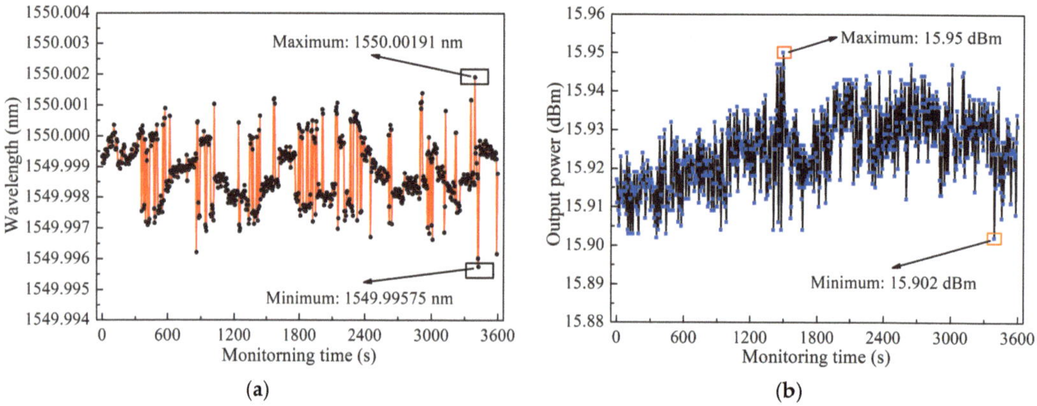

Figure 6. (**a**) Output wavelength stability curve of the tunable laser source; (**b**) output optical power stability curve of the tunable laser source.

The intrinsic linewidth of the EC-TLS is measured using an optical noise analyzer fabricated by SYCATUS. Figure 7 illustrates the relationship between the intrinsic linewidth and the resonant output wavelength. As shown in Figure 7, the intrinsic linewidth of the output laser beam remains below 10.13 kHz within the 140 nm tuning range. Notably, when the resonant output wavelength approaches 1570 nm, the intrinsic linewidth is further reduced to 5.37 kHz. This performance is attributed to the design and manufacture of the low-noise current driver and the long-term stability of the external cavity.

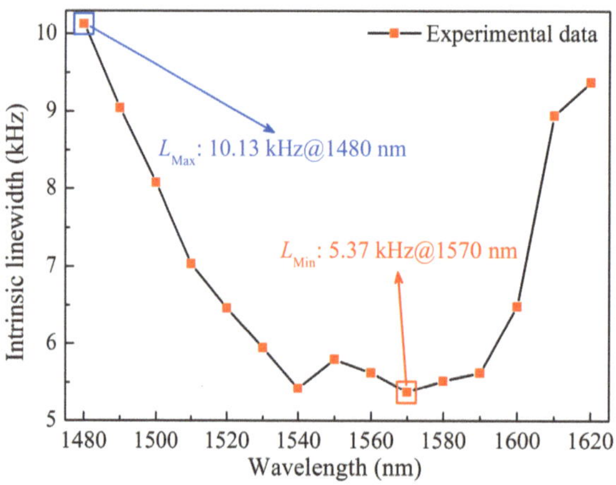

Figure 7. The relationship between the intrinsic linewidth and the resonant output wavelength.

4. Conclusions

In this paper, we have successfully constructed a wide-range, MHF tunable laser source based on external cavity feedback using a semiconductor gain chip and performed an analysis of its output characteristics. Our wavelength analysis revealed a tuning range of 140 nm without mode hopping, an RMS wavelength stability of ±0.961 pm over 1 h, and a maximum sweep speed of 200 nm/s. The optical power analysis demonstrated that the tunable laser source achieves an output power exceeding 11.14 dBm across the full C+L band, with an RMS power stability of ±0.0095 dB for over one hour at a continuous injection current of 400 mA. The spectral analysis showed that the intrinsic linewidth of the laser system is less than 10.13 kHz, and that the minimum side-mode suppression ratio is 61.65 dB. This externally coupled tunable laser source shows potential for application in high-resolution vector spectrum analysis for the discovery and understanding of various materials.

Author Contributions: Conceptualization, J.S. and L.S.; methodology, L.Q.; software, L.L.; validation, J.S. and M.P.; investigation, L.H.; resources, F.N.; data curation, Y.C.; writing—original draft preparation, J.S.; writing—review and editing, L.S. and J.H.; visualization, L.Q.; supervision, L.S., L.H., M.P., Y.C. and J.H.; project administration, F.N.; funding acquisition, L.S. All authors have read and agreed to the published version of the manuscript.

Funding: This research was funded by National Natural Science Foundation of China, grant number U23B2046; National Key R&D Program of China, grant numbers 2022YFF0705900 and 2022YFF0707102.

Institutional Review Board Statement: Not applicable.

Informed Consent Statement: Not applicable.

Data Availability Statement: Data are contained within the article.

Conflicts of Interest: The authors declare no conflicts of interest.

References

1. Maiman, T.H. Stimulated optical radiation in ruby. *Nature* **1960**, *187*, 493–494. [CrossRef]
2. Lv, Z.W.; Liu, Z.Z.; Chen, H.; Jin, D.; Hao, X.; Fan, W.Q.; Wang, Y.L.; Bai, Z.X. Review of multi-wavelength laser technology based on crystalline Raman conversion (invited). *Infrared Laser Eng.* **2023**, *52*, 20230420.
3. Shank, C.; Bjorkholm, J.; Kogelnik, H. Tunable distributed-feedback dye laser. *Appl. Phys. Lett.* **1971**, *18*, 395–396. [CrossRef]
4. Chen, H.; Cui, Y.; Li, X.; Zhang, B.; Cai, Y.; Ding, J.; Qi, Y.Y.; Yan, B.Z.; Wang, Y.L.; Lv, Z.W.; et al. High-power dual-wavelength intracavity diamond Raman laser. *Funct. Diam.* **2023**, *3*, 2282527. [CrossRef]
5. Zhang, Y.K.; Chen, H.; Bai, Z.A.; Pang, Y.J.; Wang, Y.L.; Lv, Z.W.; Bai, Z.X. Multi-wavelength red diamond Raman laser. *Infrared Laser Eng.* **2023**, *52*, 20230329.
6. Bai, Z.X.; Zhao, Z.A.; Tian, M.H.; Jin, D.; Pang, Y.J.; Li, S.S.; Yan, X.S.; Wang, Y.L.; Lv, Z.W. A comprehensive review on the development and applications of narrow-linewidth lasers. *Microwave Opt. Technol. Lett.* **2022**, *64*, 2244–2255. [CrossRef]
7. Chen, Y.F.; Tan, B.W.; Jin, D.; Chen, B.; Bai, Z.X.; Wang, K.; Wang, Y.L.; Lv, Z.W. Characteristics and suppression of beam distortion in a high repetition rate nanosecond stimulated Brillouin scattering phase conjugation mirror. *High Power Laser Sci. Eng.* **2024**, *12*, e20. [CrossRef]
8. Chen, B.; Bai, Z.X.; Zhao, G.J.; Wang, Y.L.; Lv, Z.W. Generation of high-efficiency hundred-millijoule stimulated Brillouin scattering in fused silica. *Infrared Laser Eng.* **2023**, *52*, 20230421.
9. Kasai, K.; Nakazawa, M.; Tomomatsu, Y.; Endo, T. 1.5 μm, mode-hop-free full C-band wavelength tunable laser diode with a linewidth of 8 kHz and a RIN of −130 dB/Hz and its extension to the L-band. *Opt. Express* **2017**, *25*, 22113–22124. [CrossRef]
10. Sheng, L.W.; Ge, C.L.; Cao, Q.T.; Huang, L.; Zhao, Z.A.; Li, L.F.; Qiao, S.; Zhang, A.G.; Wei, Y.; Jin, H.; et al. Wide-range external-cavity tunable semiconductor laser with mode-hopping free. *Infrared Laser Eng.* **2023**, *52*, 20230374.
11. Mroziewicz, B. External cavity wavelength tunable semiconductor lasers-a review. *Opto-Electronics Rev.* **2008**, *16*, 347–366. [CrossRef]
12. Zhang, A.G.; Qiao, S.; Sheng, L.W.; Huang, L.; Liu, Z.M.; Ju, J.W.; Zhang, Z.H.; Yin, B.Q.; Li, P.; Liu, J.Q.; et al. Study on external cavity diode laser with a wide mode-hopping free tuning range. *Front. Phys.* **2022**, *10*, 1093179. [CrossRef]
13. McGrew, W.F.; Zhang, X.; Fasano, R.J.; Schäffer, S.A.; Beloy, K.; Nicolodi, D.; Brown, R.C.; Hinkley, N.; Milani, G.; Schioppo, M.; et al. Atomic clock performance enabling geodesy below the centimetre level. *Nature* **2018**, *564*, 87–90. [CrossRef] [PubMed]
14. Schioppo, M.; Brown, R.C.; McGrew, W.F.; Hinkley, N.; Fasano, R.J.; Beloy, K.; Yonn, T.H.; Milani, G.; Nicolodi, D.; Sherman, J.A.; et al. Ultrastable optical clock with two cold-atom ensembles. *Nat. Photonics* **2017**, *11*, 5163–5166. [CrossRef]
15. Zhang, L.B.; Liu, T.; Chen, L.; Xu, G.J.; Jiang, C.H.; Liu, J.; Zhang, S.G. Development of an interference filter-stabilized external-cavity diode laser for space applications. *Photonics* **2020**, *7*, 12. [CrossRef]
16. Chi, M.J.; Jensen, O.B.; Petersen, P.M. Tuning range and output power optimization of an external-cavity GaN diode laser at 455 nm. *Appl. Optics* **2016**, *55*, 2263–2269. [CrossRef]
17. Ma, Y.R.; Yang, Q.; Tang, Y.; Chen, S.M.; Shieh, W. 1-Tb/s single-channel coherent optical OFDM transmission over 600-km SSMF fiber with subwavelength bandwidth access. *Opt. Express* **2009**, *17*, 9421–9427. [CrossRef]

18. Ding, Z.Y.; Yao, X.S.; Liu, T.G.; Du, Y.; Liu, K.; Jiang, J.F.; Meng, Z.; Chen, H.X. Compensation of laser frequency tuning nonlinearity of a long range OFDR using deskew filter. *Opt. Express* **2013**, *21*, 3826–3834. [CrossRef]
19. Leviatan, E.; Eyal, A. High resolution DAS via sinusoidal frequency scan OFDR (SFS-OFDR). *Opt. Express* **2015**, *24*, 33318–33334. [CrossRef]
20. Yin, G.L.; Lu, L.; Zhou, L.; Shao, C.; Fu, Q.J.; Zhang, J.D.; Zhu, T. Distributed directional torsion sensing based on an optical frequency domain reflectometer and a helical multicore fiber. *Opt. Express* **2020**, *28*, 16140–16150. [CrossRef]
21. Luo, M.M.; Liu, J.F.; Tang, C.J.; Wang, X.F.; Lan, T.; Kan, B.X. 0.5 mm spatial resolution distributed fiber temperature and strain sensor with position-deviation compensation based on OFDR. *Opt. Express* **2019**, *27*, 35823–35829. [CrossRef] [PubMed]
22. Paliesek, T.; Navrátil, P.; Pilař, J.; Divoký, M.; Smrž, M.; Mocek, T. Beam shaping in the high-energy kW-class laser system Bivoj at the HiLASE facility. *High Power Laser Sci. Eng.* **2023**, *11*, e79. [CrossRef]
23. Bai, Z.X.; Chen, H.; Gao, X.Q.; Li, S.S.; Qi, Y.Y.; Bai, Z.A. Highly compact nanosecond laser for space debris tracking. *Op. Materials* **2019**, *98*, 109470. [CrossRef]
24. Li, K.; Song, C.Y.; Yue, J.F.; Jia, M.Y.; Xu, Z.P.; Wu, D.; Cao, C.; Bai, Z.X.; Yu, Y.; Wang, Y.L.; et al. 500 Hz Joule-level output by sub-nanosecond Zig-Zag slab laser. *Infrared Laser Eng.* **2023**, *52*, 20230423.
25. Fedorova, K.A.; Cataluna, M.A.; Kudryashov, I.; Khalfin, V.; Rafailov, E.U. Broadly tunable InGaAsP-InP strained multi-quantum-well external cavity diode laser. *IEEE Photon. Technol. Lett.* **2010**, *22*, 1205–1207. [CrossRef]
26. Wang, Y.; Wu, H.; Chen, C.; Zhou, Y.L.; Wang, Y.B.; Liang, L.; Tian, Z.H.; Qin, L.; Wang, L.J. An ultra-high-SMSR external-cavity diode laser with a wide tunable range around 1550 nm. *Appl. Sci.* **2019**, *9*, 4390. [CrossRef]
27. Gong, H.; Liu, Z.G.; Zhou, Y.L.; Zhang, W.B. Extending the mode-hop-free tuning range of an external-cavity diode laser by synchronous tuning with mode matching. *Appl. Opt.* **2014**, *53*, 7878–7884. [CrossRef] [PubMed]
28. Zhu, J.Y.; Qiao, D.; Jones, A.; Zhang, B.; Li, K.; Copner, N. 1.7 THz tuning range pivot-point-independent mode-hop-free external cavity diode laser. *Opt. Express* **2023**, *31*, 3970–3983. [CrossRef]
29. Dong, J.T.; Chen, Z.W.; Sheng, L.W.; Huang, L. Collimation system of external cavity tunable diode lasers (invited). *Electro-Opt. Technol. Appl.* **2024**, *39*, 56–59.

photonics

MDPI

Article

A 2 μm Gallium Antimonide Semiconductor Laser Based on Slanted, Wedge-Shaped Microlens Fiber Coupling

Zhaohong Liu [1,2,*], Jiayue Wang [1], Ning Li [1,3], Zhongwei Yang [1], Shaowen Li [1], Sensen Li [1,2,3], Wei Wang [4], Heshig Bayan [4], Weining Cheng [3], Yu Zhang [2], Zhuokun Wu [3], Hongyu Sun [3], Yuanqing Xia [1], Yulei Wang [1] and Zhiwei Lu [1]

1 Hebei Key Laboratory of Advanced Laser Technology and Equipment, Tianjin 300401, China; 202231903042@stu.hebut.edu.cn (J.W.); 202331905043@stu.hebut.edu.cn (N.L.); 202221901035@stu.hebut.edu.cn (Z.Y.); 173757@stu.hebut.edu.cn (S.L.); sensli@163.com (S.L.); xiayq@hebut.edu.cn (Y.X.); wyl@hebut.edu.cn (Y.W.); zhiweilv@hebut.edu.cn (Z.L.)

2 Shanxi Key Laboratory of Advanced Semiconductor Optoelectronic Devices and Integrated System, Jincheng 048000, China; zhangyu@semi.ac.cn

3 China Electronics Technology Optoelectronic Research Institute, Tianjin 300308, China; 202331905008@stu.hebut.edu.cn (Z.W.); 202331905065@stu.hebut.edu.cn (H.S.)

4 Changchun Institute of Optics, Fine Mechanics and Physics, Chinese Academy of Sciences, Changchun 130033, China; 202321901049@stu.hebut.edu.cn (W.W.); bayin888@sina.com (H.B.)

* Correspondence: lzh@hebut.edu.cn

Abstract: Semiconductor lasers with a wavelength of 2 μm, composed of antimonide materials, find important applications in trace gas detection, laser medicine, and free-space optical communication, among others. In this paper, a more suitable microlens shape for 2 μm gallium antimonide semiconductor lasers is designed. Based on the fiber coupling efficiency model, the parameters of the designed slanting wedge-shaped microlens fiber are optimized to improve laser beam quality. The large tangent angle on both sides of the slanted, wedge-shaped microlens fiber is calculated using Snell's law, and the fiber core diameter and small wedge angle are determined through space fiber coupling experiments. After packaging the fiber coupling module with the chip, the laser output beam exhibits good overall symmetry in the spot with a uniform intensity distribution. The maximum output power is approximately 210 mW, demonstrating good power stability.

Keywords: antimonide semiconductor laser; 2 μm IR; fiber coupled; microlens

1. Introduction

Semiconductor lasers utilizing antimonide materials in the mid-infrared spectrum exhibit multiple gas molecule absorption peaks within the 2 μm band. Additionally, they exploit an atmospheric window characterized by high transmittance [1,2], rendering these lasers pivotal in the realms of trace gas detection, laser therapy, materials processing, and free-space optical communication [3–6]. Since the 19th century, extensive research efforts have been dedicated to mid-infrared semiconductor lasers, with a particular emphasis on antimonide-based semiconductor materials. These materials offer indispensable alloys, bandgaps, and bandstructures, constituting fundamental technologies in the contemporary landscape of mid-infrared semiconductor lasers [7–9]. Nevertheless, the current state of antimonide semiconductor lasers is encumbered by limitations in power output and beam quality, thereby constraining their potential applications.

In this paper, we employ the microlens fiber coupling technique to shape the laser beam, aiming to enhance coupling efficiency and improve beam quality. Based on different optical structures, coupling methods can be categorized into combination lens fiber coupling and microlens fiber coupling. Microlens fiber coupling involves the direct processing or attachment of a specific microlens to the end face of the fiber [10], enabling direct coupling with the semiconductor laser. The microlens shape can typically be conical [11], hyperbolic [12], parabolic [13], cylindrical [14], and more. This paper utilizes the microlens fiber coupling method, directly coupling the microlens fiber with the light source. This approach offers several advantages, including a simple and compact structure, ease of alignment with the semiconductor laser chip, simple fabrication, and low production cost. A study on coupling a microlens fiber to a semiconductor laser shares similarities with the external combined lens fiber coupling scheme. The lens structure significantly influences whether high coupling efficiency can be achieved. Some representative work in related areas is presented below. In 2010, Yang et al. from Yishou University, Taiwan, compared the coupling efficiencies of tapered hyperbolic microlens fibers and tapered hemispherical microlens fibers and found that tapered hyperbolic microlens have much higher coupling efficiencies [15]. In 2014, Das from India studied the coupling efficiency of hyperbolic microlens fibers at two different wavelengths, 1.3 μm and 1.5 μm, using the ABCD matrix and obtained a higher coupling efficiency at 1.3 μm [12]. In 2015, Sanker at the Cape Institute of Technology, India, used a new type of fiber with an inverted-taper microlens at the fiber tip in order to maximize the coupling efficiency between a semiconductor laser and a single-mode fiber [16]. In addition, researchers have proposed a thermally expandable core fiber structure with a high numerical aperture to improve the coupling efficiency by using the thermally expandable core technique [17,18]. The thermal expansion core scheme offers higher coupling efficiency and greater lateral and longitudinal tolerances, but it has a lower tilt tolerance than conventional single-mode fiber schemes. In 2022, Nie et al. from the University of Wollongong, Australia, built a sensing system by coupling a semiconductor laser with optical feedback to a fiber [19]. The system can realize remote distance sensing, which is important for the practical application of semiconductor lasers coupled with optical fibers. At present, the development and application of microlens fiber coupling technology [20–23] are still hindered by challenges such as low coupling efficiency and a small mounting tolerance rate. These issues impede the effective resolution of problems related to the low power and poor beam quality of antimonide semiconductor lasers [24–28]. Therefore, it is necessary to analyze and study the microlens fiber end-face structure that is better suited for these lasers.

In this paper, we conducted a simulation study using ZEMAX OpticStudio2019 software to analyze the fiber coupling between common microlens fiber structures and antimonide semiconductor laser sources. The study aimed to compare coupling efficiency and mounting tolerance among different microlens fiber structures. Combining the beam characteristics of 2 μm antimonide semiconductor lasers with the advantages of wedge microlens and tapered microlens, we introduced a novel end-face structure named a slanted, wedge-shaped microlens fiber. This fiber demonstrated higher coupling efficiency, a larger modulation tolerance, and effectively improved the laser beam quality. The laser achieved a stable power output of 210 mW, with the output power showing increased stability near a current of 1.5 A. The RMS stability and peak-to-peak stability of the laser were 0.39% and 2.17%, respectively, during continuous output at a current of 1.5 A over two hours.

2. Theory and Simulation

The main common microlens fiber shapes include planar fibers, wedge microlens fibers, and tapered microlens fibers. Due to the large divergence angle of the semiconductor laser, direct coupling of a planar fiber to the light source results in poor coupling efficiency and low fiber mounting tolerance if the fiber is not processed. In this chapter, we simulate the coupling efficiency and mounting error rate of wedge and tapered microlens fibers. By combining the semiconductor laser chip used in this paper with the characteristics of wedge and tapered microlens fibers, we designed the slanted wedge microlens fiber. The structured fiber achieves a simulation coupling efficiency of 78.5% with uncoated end faces and significantly improved mounting tolerance.

In this paper, the M-1940-0500-A antimonide semiconductor laser chip, previously researched in the Superlattice Laboratory of the Institute of Semiconductor Research [29] at the Chinese Academy of Sciences, serves as a prototype design simulation light source. A semiconductor laser light source is constructed based on chip parameters using ZEMAX sequence mode. Specific parameters of the optical fiber are configured in ZEMAX. The solid model of the fiber optic microlens is then established in SolidWorks 3D CAD2020 software. Subsequently, the non-sequential component in ZEMAX is utilized to import the microlens model, adjust its position, and nest it in the fiber optic end face to obtain the coupled model diagram of the microlens fiber. The geometric image analysis function is employed to calculate the coupling efficiency of the fiber, considering Fresnel reflection loss and optical polarization generated at the air-fiber split interface during the calculation process.

Typical angles for tapered microlens range from 45° to 100°. Microlens fibers with different taper angles can be selected based on the type of laser chip and desired light output performance. Taking the example of a 100° tapered microlens fiber, through simulation and analysis in ZEMAX, the coupling efficiency between the semiconductor laser and the tapered microlens fiber is determined to be 72.3%. Figure 1 displays the NSC solid model diagram of the optical fiber and illustrates the effects of positional and rotational degree errors on the fiber coupling efficiency. In Figure 1b, the mounting tolerances along the X and Y axes for the tapered microlens fiber are small, while the mounting tolerance along the Z-axis is relatively large. Figure 1c indicates that changes in the rotation angle of the fiber around its own axis do not affect the coupling efficiency of the tapered microlens fiber. This is primarily attributed to the circularly symmetric shape of the end face of the tapered microlens fiber, causing the rotation of the fiber around its own axis to have no impact on the light field pattern of the end face.

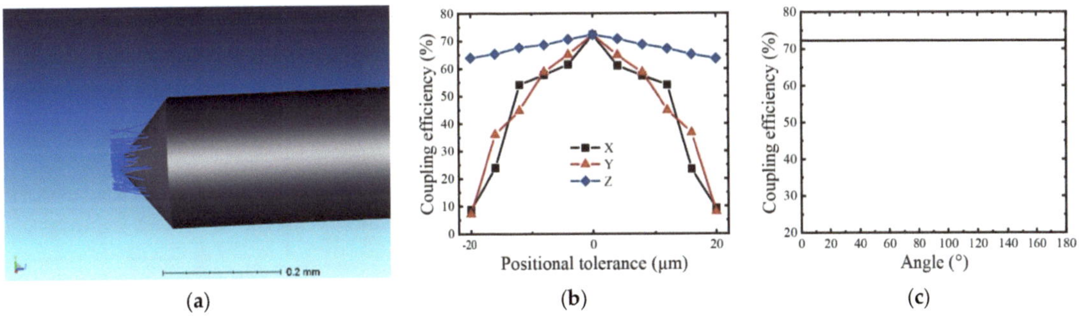

Figure 1. (**a**) Diagram of the NSC solid model of a tapered microlens fiber; (**b**) effect of tapered fiber position error on fiber coupling efficiency; and (**c**) effect of tapered fiber rotational degree error on fiber coupling efficiency.

For wedge-shaped microlens fibers, the size of the wedge angle has a crucial effect on the coupling efficiency of the fiber. We calculate the size of the wedge angle of the wedge microlens fiber according to Snell's law. Let the half wedge angle of the wedge lens be θ_x and the vertical divergence angle of the semiconductor laser chip be $\theta_\perp$, then we have the following relation:

$$\theta_\chi = \theta_\perp + \theta_1 \tag{1}$$

The following is known from Snell's law:

$$n_0 \sin\alpha = n_1 \sin\beta \tag{2}$$

Where n_0 is the refractive index of air; n_1 is the refractive index of the fiber core.

From practical application, it is necessary to make the incident light parallel to the z-axis direction, so that the following can be derived:

$$\sin\alpha = \sin\left(\frac{\pi}{2} - \theta_1\right) = \cos\theta_1 = n_1 \sin\beta = n_1 \sin\left(\frac{\pi}{2} - \theta_x\right) = n_1 \cos\theta_x \tag{3}$$

Bringing Equation (1) into Equation (3) yields the following:

$$\cos(\theta_x - \theta_\perp) = n_1 \cos\theta_x \quad (4)$$

The relation between the half-wedge angle θ_x of the wedge microlens fiber and the vertical divergence angle $\theta_\perp$ of the semiconductor laser is obtained after rationalization and simplification:

$$\theta_x = \tan^{-1}\left(\frac{n_1 - \cos\theta_\perp}{\sin\theta_\perp}\right) \quad (5)$$

The semiconductor laser chip utilized in this paper is known to have a vertical divergence angle of 50°, and the wedge angle of the wedge microlens fiber is determined to be approximately 70° according to Equation (5). Figure 2 illustrates the impact of position error and rotation error on fiber coupling efficiency for a 70° wedge microlens fiber. The highest coupling efficiency for the wedge microlens fiber, obtained through the optimization function of ZEMAX software, is recorded at 83.7%. In Figure 2b, the Y-axis tuning tolerance of the wedge microlens fiber demonstrates a significant improvement, and the Z-axis fiber coupling efficiency changes smoothly, while the X-axis tuning tolerance remains relatively small. In Figure 2c, it is evident that the rotation angle error of the wedge microlens fiber has a substantial impact on the coupling efficiency. This effect is attributed to the significant structural differences between the wedge microlens in the transverse and longitudinal axis directions. Any rotation introduces a mode-matching mismatch at the fiber end face, resulting in a pronounced cliff-type decrease in coupling efficiency.

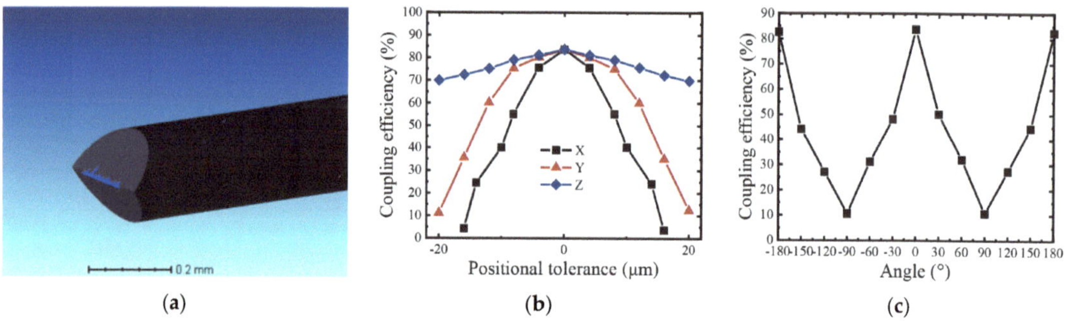

Figure 2. (**a**) Diagram of the NSC solid model of a wedge-shaped microlens fiber; (**b**) effect of wedge-shaped fiber position error on fiber coupling efficiency; and (**c**) effect of wedge-shaped fiber rotational degree error on fiber coupling efficiency.

Combining the findings from the above research and discussions, it is observed that the coupling efficiency is higher when compared with both the wedge microlens fiber and the tapered microlens fiber. However, the tuning tolerance of the wedge fiber is relatively small, especially when the fiber rotates around its own axis, leading to a significant impact on the coupling efficiency. This limitation is not conducive to practical experiments and tuning. On the other hand, although the tapered microlens fiber is not affected by the rotation angle, its coupling efficiency is lower. In response to these considerations, this paper proposes a new end-face structure called the slanted wedge microlens fiber, which is designed by combining the advantages of wedge microlens and conical microlens.

The end face of a slanted wedge microlens fiber is conceptualized as a slant cut on top of a wedge microlens, forming a small bevel to augment the numerical aperture of the fiber and enhance its light-harvesting capability. Consequently, the large tangent angle on both sides of the slanted, wedge-shaped microlens fiber is identical to the wedge angle of the wedge-shaped microlens fiber, measuring 70°. Utilizing the genetic algorithm optimization function in computer software, the small tangent angle of the slanted, wedge-shaped microlens fiber is optimized. Figure 3 illustrates the impact of changes in the small tangent angle on fiber coupling efficiency, with a fixed large tangent angle of 70° on both sides. The optimization process reveals that the maximum fiber coupling efficiency occurs when the small tangent angle is approximately 28°. Through the multi-mode fiber coupling simulation function of ZEMAX, it is determined that the coupling efficiency of the slanted, wedge-shaped microlens fiber can reach 78.5% without coating at the end face, as analyzed through geometric image analysis in Optic Studio.

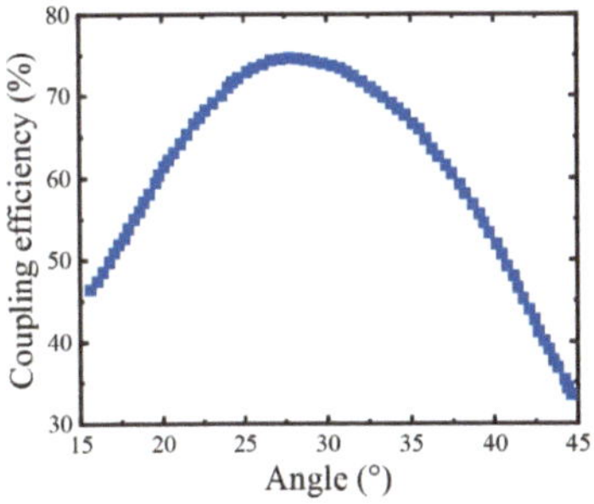

Figure 3. Fiber coupling efficiency versus small wedge angle of slanted wedge microlens fiber.

Figure 4 presents the NSC solid model diagram of the slanted wedge microlens fiber, along with a plot depicting the impact of position error and rotation degree error on fiber coupling efficiency. From Figure 4b, it is evident that the X-axis and Y-axis mounting tolerances of the optical fiber have been significantly enhanced, resulting in close adjustment tolerances in both transverse and longitudinal directions. The adjustment tolerance in the Z-axis direction is also improved compared to wedge microlens fibers, attributable to the increased numerical aperture that enhances light harvesting capability. In Figure 4c, the adjustment tolerance of the fiber in rotation around its own axis is improved compared to that of the wedge microlens fiber. When compared to the tapered microlens fiber, the adjustment tolerance between 0° and 40° is relatively large, and further improvement is needed in the adjustment tolerance between 60° and 90°.

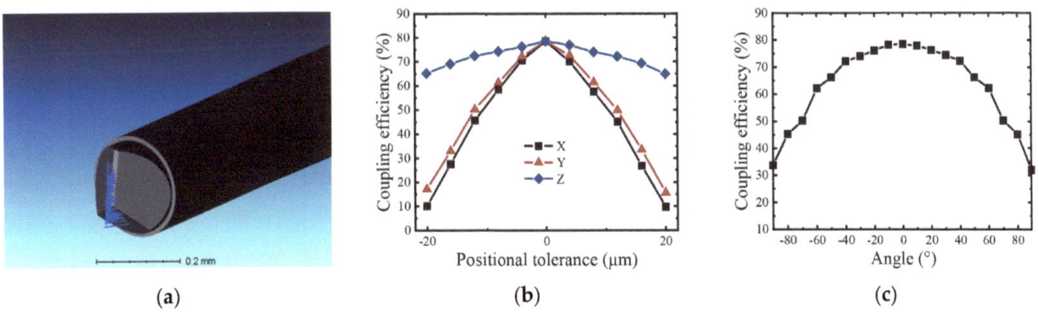

Figure 4. (**a**) Diagram of the NSC solid model of a slanted wedge microlens fiber; (**b**) effect of slanted wedge fiber position error on fiber coupling efficiency; and (**c**) effect of slanted wedge fiber rotational degree error on fiber coupling efficiency.

3. Experiment

Optical System

Firstly, we performed the spatial fiber coupling experiments on the slanted, wedge-shaped microlens fiber. Different small wedge angles of 105 μm and 200 μm of the slanted, wedge-shaped microlens fiber (no permeability-enhancing film on the microlens surface) were coupled to the semiconductor laser chip, respectively, and the large wedge angle on both sides of the slanted, wedge-shaped microlens fiber is unchanged at 70°, and the results are shown in Figure 5. It can be seen from the figure that the coupling efficiency is 31.2%, 34.6%, and 29.3% for the small wedge angle of 20°, 30°, and 40° for 105 μm core diameter, respectively, and 51.2%, 52.5%, and 50.3% for the small wedge angle of 20°, 30°, and 40° for 200 μm core diameter, respectively. It can be seen that the coupling efficiency is higher for 105 μm and 200 μm core diameters when the small wedge angle of the slanted wedge microlens fiber is 30°. The coupling efficiency of the fiber is greatly improved when the core diameter is expanded from 105 μm to 200 μm, with 64.1%, 51.7%, and 71.7% improvement when the small wedge angle of the slanted wedge microlens fiber is 20°, 30°, and 40°, respectively. The above data were measured at a current of 1.5 A and a voltage of 1.85 V. From the experimental results, it is obvious that the 200 μm slanted wedge fiber with a small wedge angle of 30° has more advantages in coupling with the semiconductor laser chip.

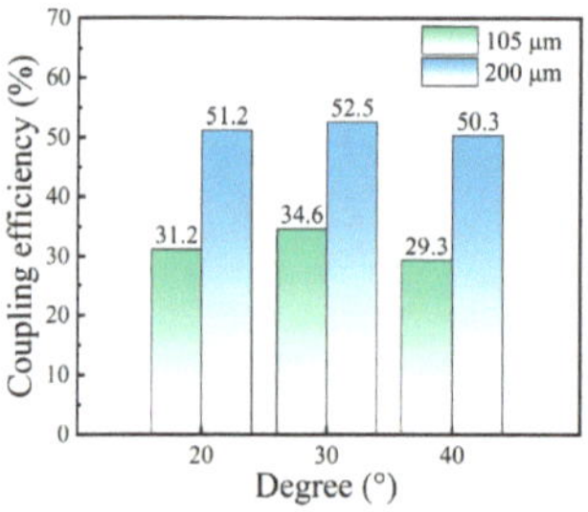

Figure 5. Experimental results of fiber coupling with slanted, wedge-shaped microlens with different small wedge angles of 105 μm and 200 μm.

Spatial fiber coupling experiments were performed using microlens fibers with wedge, tapered, and slanted wedge end faces. Figure 6a shows the physical and SEM images of the three microlens fibers. The coupling results of the three microlens fibers with a wedge angle of 70°, a taper angle of 80°, and a small wedge angle of 30° were put together for comparison, as shown in Figure 6b.

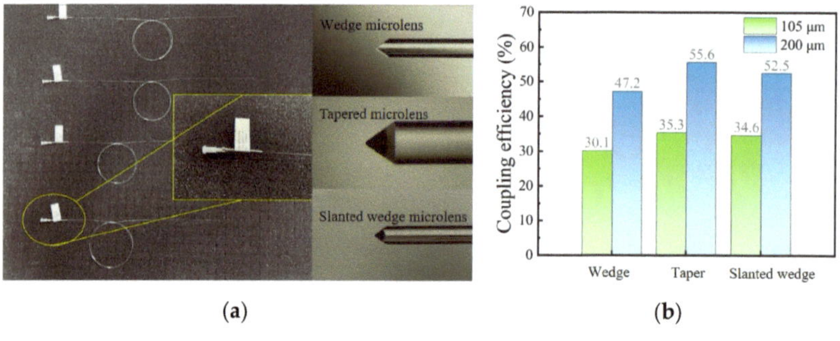

Figure 6. (**a**) Physical drawings of three optical fibers and microscopic microlens shapes of the fiber end faces; (**b**) comparison of coupling efficiencies of 105 μm and 200 μm wedge, taper, and slant-wedge end-face microlens fibers.

For the fiber coupling experiments, a 200/220 type numerical aperture of 0.22, a small wedge angle of 30° slanted, wedge-shaped microlens fiber, and a semiconductor laser chip are enclosed in a metal tube housing. The experiments involve fixing the fiber at the end to achieve maximum power, dispensing a fixed fiber, and finally sealing the setup to form a chip-fiber coupling module. Following the coupling between the semiconductor laser chip and the microlens fiber in the housing, the measured coupling efficiency exceeds 50%, meeting the anticipated standard.

Subsequently, after encapsulating the water-cooled module, the laser is assembled with the circuit module. The water-cooled module and circuitry are enclosed in the designed laser housing, secured with screws. Figure 7 illustrates the optical path of the gallium antimonide semiconductor laser. The core part of the semiconductor laser chip is the PN junction, which is dissected along the natural crystal surface to form two smooth end surfaces known as solvation surfaces. These surfaces function as planar reflectors, establishing a resonant cavity.

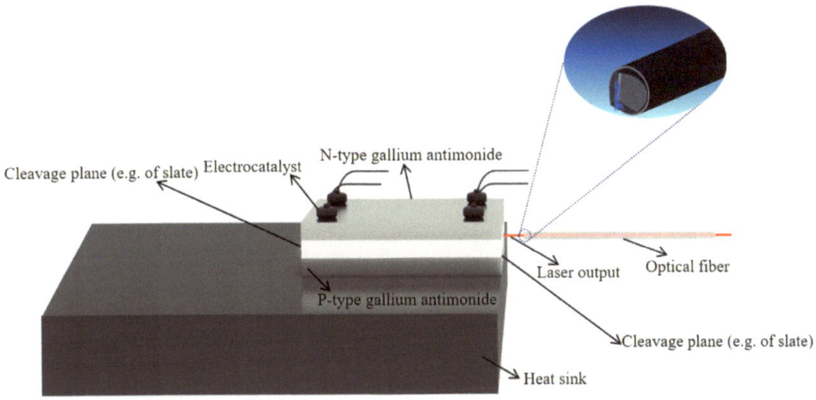

Figure 7. Gallium antimonide semiconductor laser optical path diagram.

Upon constructing the gallium antimonide semiconductor laser, a beam test is conducted to verify the anticipated results. In the performance test experiments, the focusing lens has a focal length of 500 mm. The beam quality analyzer is a CinCam product produced by CINOGY, Göttingen, Germany. The power meter head is a Vega handheld power meter manufactured by OPHIR, Israel, with the detector being a model 3A-P probe from the same company capable of detecting a power range of 15 μW–3 W. Additionally, the spectrometer used is a 771 series laser spectrum analyzer produced by British BRISTOL, featuring a spectral resolution as high as 2 GHz, a wavelength accuracy of ±0.0001 nm, and an optical rejection ratio exceeding 40 dB. The experimental test platform is completed by connecting the water cooler, gallium antimonide semiconductor laser, focusing lens, beam quality analyzer (power meter/spectrometer), and a laptop computer in the specified order.

4. Results and Discussion

Upon completing the construction of the platform, a comparison is made between the performance of the laser after fiber coupling and the performance of the chip when the light comes out directly. Figure 8 displays the PIV curves of the chip and the semiconductor laser.

In Figure 8a, the PIV plot of the chip reveals a threshold current of 0.3 A and a maximum output power of approximately 345 mW. The calculated average slope efficiency (W/A) for the chip, based on the measured data, is 0.17. It is observed that the power of the chip is more stable near the current of 1.5 A.

Figure 8b depicts the PIV curve of the gallium antimonide semiconductor laser, showing a threshold current of 0.2 A and a maximum output power of about 210 mW. The calculated average slope efficiency (W/A) for the semiconductor laser, based on the measured data, is 0.09. Similar to the chip, the power of the semiconductor laser is more stable in the vicinity of a current of 1.5 A.

A comparison with the chip test results reveals that the threshold current of the laser is reduced, attributed to the improved cooling effect of water cooling from an external water tank compared to natural convection cooling by air. This reduction in threshold current is a result of the enhanced cooling efficiency. Additionally, the power of the laser is found to be diminished compared to the direct output power of the chip, owing to losses incurred during the fiber coupling and transmission process.

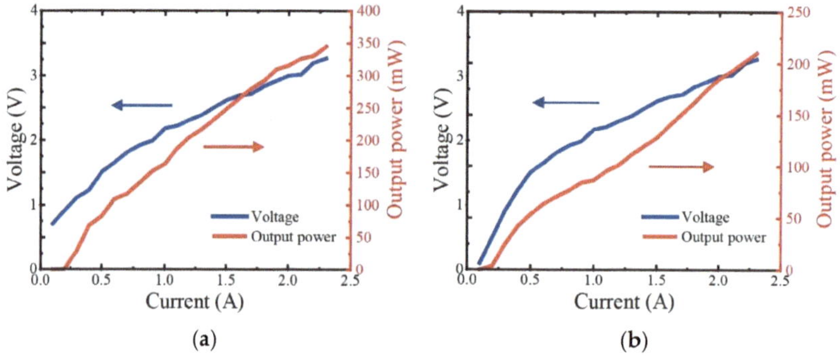

Figure 8. PIV curves of a chip and a semiconductor laser: (**a**) PIV curve of a semiconductor laser chip; (**b**) PIV curve of a gallium antimonide semiconductor laser.

Figure 9a displays the spectra of a gallium antimonide semiconductor laser at a current of 1.5 A. At this juncture, the primary wavelength of the semiconductor laser is observed to be 1935 nm, and the spectrogram exhibits a dendritic spectral distribution with a line width of approximately 15 nm. In Figure 9b, a comparison of spectrograms for gallium antimonide semiconductor lasers at different operating currents is presented. The spectral range of the semiconductor laser spans from 1910 nm to 1960 nm, with a gradual increase in current. Simultaneously, it is evident that the semiconductor laser output beam, under the influence of different currents, exhibits multiple longitudinal modes. This phenomenon is attributed to the laser chip for the FP cavity structure, where the frequency selection is not enough.

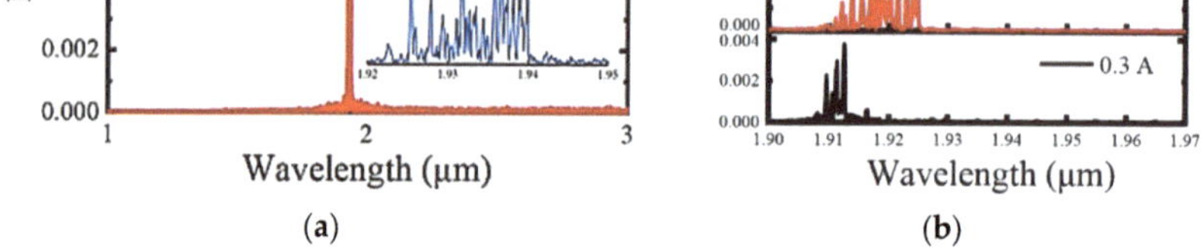

Figure 9. (**a**) Spectrogram of gallium antimonide semiconductor laser; (**b**) normalized spectra of gallium antimonide semiconductor laser at different operating currents.

The beam quality of the gallium antimonide semiconductor laser and the laser chip was subsequently tested using a beam quality analyzer produced by CINOGY. Figure 10 displays the output beam spot images of the chip and the gallium antimonide semiconductor laser at 1.5 A. In Figure 10a, the output beam spot of the chip is observed to be elliptical, with a relatively large spot length and width. The fast-axis and slow-axis divergence angles of the semiconductor chip are calculated to be 50° and 30°, respectively. Figure 10b illustrates the output beam spot of the laser, which is circular with good overall symmetry and a Gaussian distribution featuring uniform intensity. The output beam spot of the laser is notably superior to that of the chip, indicating a significant improvement in beam quality.

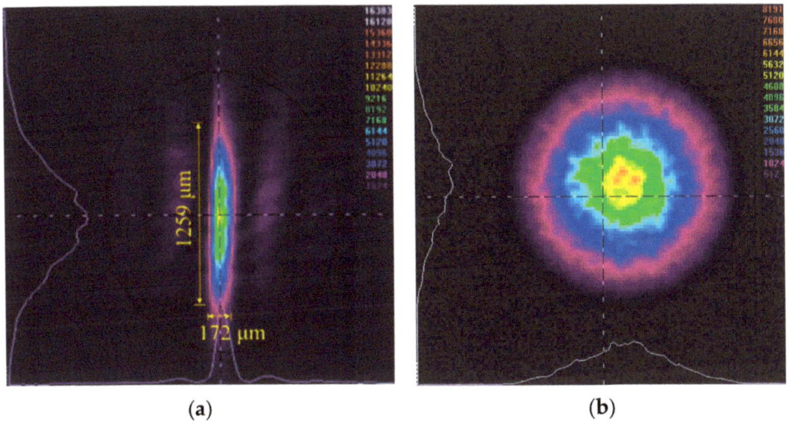

Figure 10. Output spot image: (**a**) output spot image of the chip; (**b**) output spot image of the gallium antimonide semiconductor laser.

For semiconductor lasers, the beam quality factor M^2 is ideally 1. In practice, the smaller the value, i.e., the closer to 1, the better the beam quality; conversely, the poorer the beam quality. Figure 11 shows the near-field spot images of a gallium antimonide semiconductor laser with an output from 100 mm to 500 mm at 1.5 A current measured using a beam quality analyzer, and the XY-axis spot size variation curve of the laser is obtained via data fitting. The output power of the laser is 128 mw at a current of 1.5 A. The beam quality factors in the XY direction are $MX^2 = 1.30$ and $MY^2 = 1.42$, and the divergence angles of the laser in the fast and slow axes are 35°and 28°, respectively, which shows that the beam quality of the semiconductor laser after fiber coupling has been greatly improved.

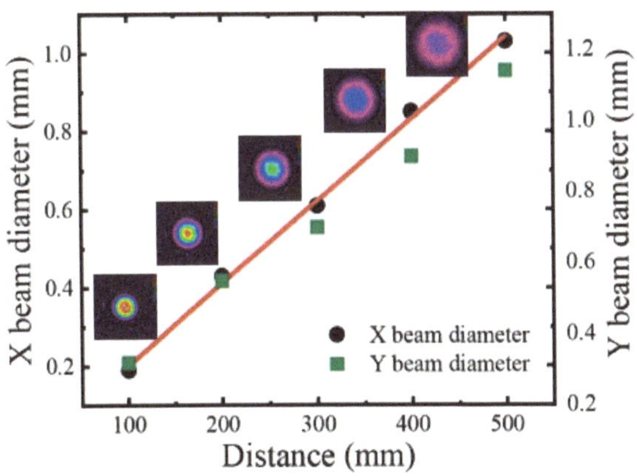

Figure 11. Relationship between output spot size and transmission distance of semiconductor lasers.

Figure 12 shows the corresponding power stability curve of the gallium antimonide semiconductor laser in the case of two hours of continuous output at a current of 1.5 A. Tested at an ambient temperature of 20° and an ambient humidity of 40% (no condensation). According to the results, the RMS stability and peak-to-peak stability of the gallium antimonide semiconductor laser can be calculated to be 0.39% and 2.17%, respectively.

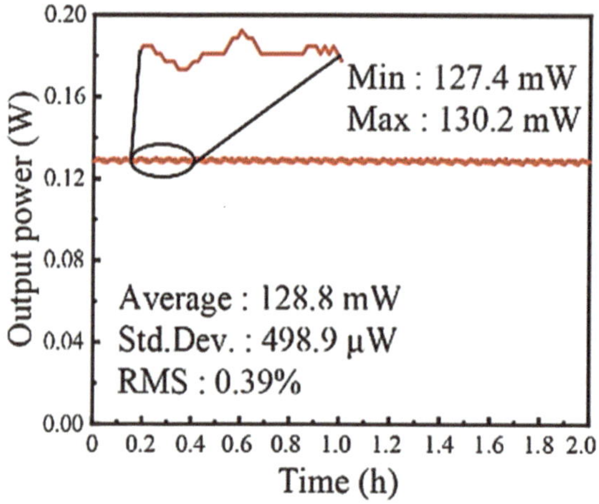

Figure 12. Power stability curve of gallium antimonide semiconductor laser.

5. Conclusions

In this paper, we calculated the angle of 70° for the large tangent angle on both sides of the slanted wedge microlens fiber according to Snell's law, and in the spatial fiber coupling of the slanted wedge microlens fiber, we found that the 200 μm slanted wedge fiber with a small wedge angle of 30° has obvious advantages when coupling with a semiconductor laser chip. We have used a 200/220 type optical fiber with a numerical aperture of 0.22 and a small wedge angle of 30° to complete the fiber coupling with a semiconductor laser chip in a metal tube housing. In this paper, we also completed the construction of a gallium antimonide semiconductor laser and performed a performance analysis of the laser. During the analysis of beam quality, we found that the diagonal wedge microlens fiber coupling effectively improved the beam quality of the laser. Analyzing the output power of the laser, we found that the threshold current of the semiconductor laser is 0.2 A and the maximum output power is about 210 mW. The RMS stability and peak-to-peak stability of this laser were 0.39% and 2.17%, respectively, under two hours of continuous output at 1.5 A current.

Author Contributions: Conceptualization, J.W.; methodology, J.W. and N.L.; software, Z.Y.; validation, S.L. (Shaowen Li); investigation, W.W.; resources, H.B.; data curation, W.C.; writing—original draft preparation, J.W.; writing—review and editing, Z.L. (Zhaohong Liu) and J.W.; visualization, Y.Z.; supervision, Z.L. (Zhaohong Liu), S.L. (Sensen Li), Z.W., H.S., Y.X., Y.W. and Z.L. (Zhiwei Lu); project administration, Z.L. (Zhaohong Liu); funding acquisition, Z.L. (Zhaohong Liu). All authors have read and agreed to the published version of the manuscript.

Funding: This research is funded by National Natural Science Foundation of China (61905064, 61975050), the China Postdoctoral Science Foundation (300428), the Hebei Province Postdoctoral special grant (B2022005003), the Research Projects of High Education Institutions of Hebei Province (QN2019201), the Natural Science Research Foundation of Hebei University of Technology (JBKYXX2002), the Open Project Program of Shanxi Key Laboratory of Advanced Semiconductor Optoelectronic Devices and Integrated (2022SZKF04), and the National Key R&D Program of China (2022YFB3606100, 9813E2301X04).

Institutional Review Board Statement: Not applicable.

Informed Consent Statement: Not applicable.

Data Availability Statement: Data are contained within the article.

Conflicts of Interest: The authors declare no conflicts of interest.

References

1. Werle, P. A review of recent advances in semiconductor laser based gas monitors. *Spectrochim. Acta Part A. Mol. Biomol. Spectrosc.* **1998**, *54*, 197–236. [CrossRef]
2. Yin, Z.; Tang, X. A review of energy bandgap engineering in III-V semiconductor alloys for mid-infrared laser applications. *Solid State Electron.* **2007**, *51*, 6–15. [CrossRef]
3. Garnache, A.; Kachanov, A.A.; Stoeckel, F.; Planel, R. High-sensitivity intracavity laser absorption spectroscopy with vertical-external-cavity surface-emitting semiconductor lasers. *Opt. Lett.* **1999**, *24*, 826–828. [CrossRef]
4. López-Lorente, Á.I.; Mizaikoff, B. Mid-infrared spectroscopy for protein analysis: Potential and challenges. *Anal. Bioanal. Chem.* **2016**, *408*, 2875–2889. [CrossRef]
5. Roth, S.; Nakata, Y.; Neuenschwander, B.; Xu, X.; Voisiat, B.; Gaponov, D.; GečYs, P.; Lavoute, L.; Silva, M.; Hideur, A. Material processing with ultra-short pulse lasers working in 2μm wavelength range. In *Laser Applications in Microelectronic and Optoelectronic Manufacturing (LAMOM) XX*; SPIE: Amsterdam, The Netherlands, 2015; p. 935014.
6. Ducanchez, A.; Cerutti, L.; Grech, P.; Genty, F.; Tournie, E. Mid-infrared Gasb-based Ep-vcsel Emitting At 2.63 μm. *Electronicsletters* **2009**, *45*, 265–267.
7. Tournié, E.; Cerutti, L. *Mid-Infrared Optoelectronics: Materials, Devices, and Applications*; Woodhead Publishing: Sawston, UK, 2019.
8. Helkey, R.; Saleh, A.A.; Buckwalter, J.; Bowers, J.E. High-performance photonic integrated circuits on silicon. *IEEE J. Sel. Top. Quantum Electron.* **2019**, *25*, 1–15. [CrossRef]
9. Ramirez, J.M.; Elfaiki, H.; Verolet, T.; Besancon, C.; Achouche, M. III-V-on-Silicon Integration: From Hybrid Devices to Heterogeneous Photonic Integrated Circuits. *IEEE J. Sel. Top. Quantum Electron.* **2019**, *26*, 2939503. [CrossRef]
10. Yeh, S.-M.; Huang, S.-Y.; Cheng, W.-H. A new scheme of conical-wedge-shaped fiber endface for coupling between high-power laser diodes and single-mode fibers. *J. Light. Technol.* **2005**, *23*, 1781.
11. Yeh, S.-M.; Lu, Y.-K.; Huang, S.-Y.; Lin, H.-H.; Hsieh, C.-H.; Cheng, W.-H. A novel scheme of lensed fiber employing a quadrangular-pyramid-shaped fiber endface for coupling between high-power laser diodes and single-mode fibers. *J. Light. Technol.* **2004**, *22*, 1374. [CrossRef]
12. Das, B.; Maiti, A.K.; Gangopadhyay, S. Laser diode to single-mode circular core dispersion-shifted/dispersion-flattened fiber excitation via hyperbolic microlens on the fiber tip: Prediction of coupling efficiency by ABCD matrix formalism. *Optik* **2014**, *125*, 3277–3282. [CrossRef]
13. Yu, Y.; Dou, W. ABCD matrix for reflection and refraction of Gaussian beams on the interface of an elliptic paraboloid. *J. Infrared Millim. Terahertz Waves* **2010**, *31*, 1304–1311. [CrossRef]
14. Xu, H.; Zhou, H.; Duan, J.-A. Allowable aperture considerations for laser diode coupling to cylindrical lensed fiber: Efficiency computation with ABCD matrix. *Optik* **2019**, *185*, 614–619. [CrossRef]
15. Yang, H.M.; Chen, C.T.; Ro, R.; Liang, T.C. Investigation of the efficient coupling between a highly elliptical Gaussian profile output from a laser diode and a single mode fiber using a hyperbolic-shaped microlens. *Opt. Laser Technol.* **2010**, *42*, 918–926. [CrossRef]
16. Perumal Sankar, S.; Hariharan, N.; Varatharajan, R. A novel method to increase the coupling efficiency of laser to single mode fibre. *Wirel. Pers. Commun.* **2016**, *87*, 419–430. [CrossRef]
17. Kliros, G.S.; Divari, P.C. Coupling characteristics of laser diodes to high numerical aperture thermally expanded core fibers. *J. Mater. Sci. Mater. Electron.* **2009**, *20*, 59–62. [CrossRef]
18. Kliros, G.; Tsironikos, N. Variational analysis of propagation characteristics in thermally diffused expanded core fibers. *Optik* **2005**, *116*, 365–374. [CrossRef]
19. Nie, B.; Ruan, Y.; Yu, Y.; Guo, Q.; Fang, C.; Xi, J.; Tong, J.; Du, H. Achieving long distance sensing using semiconductor laser with optical feedback by operating at switching status. *Sensors* **2022**, *22*, 963. [CrossRef]
20. Shterengas, L.; Kipshidze, G.; Hosoda, T.; Liang, R.; Feng, T.; Wang, M.; Stein, A.; Belenky, G. Cascade pumping of 1.9–3.3 μm type-i quantum well GaSb-based diode lasers. *IEEE J. Sel. Top. Quantum Electron.* **2017**, *23*, 1–8. [CrossRef]
21. Shang, C.; Hughes, E.; Wan, Y.; Dumont, M.; Koscica, R.; Selvidge, J.; Herrick, R.; Gossard, A.C.; Mukherjee, K.; Bowers, J.E. High-temperature reliable quantum-dot lasers on Si with misfit and threading dislocation filters. *Optica* **2021**, *8*, 749–754. [CrossRef]
22. Monge-Bartolome, L.; Cerba, T.; Díaz-Thomas, D.A.; Bahriz, M.; Calvo, M.R.; Boissier, G.; Baron, T.; Rodriguez, J.-B.; Cerutti, L.; Tournié, E. Etched-cavity GaSb laser diodes on a MOVPE GaSb-on-Si template. *Opt. Express* **2020**, *28*, 20785–20793. [CrossRef]
23. Zhou, H.; Xu, H.; Duan, J.-a. Review of the technology of a single mode fiber coupling to a laser diode. *Opt. Fiber Technol.* **2020**, *55*, 102097. [CrossRef]
24. Spott, A.; Stanton, E.J.; Torres, A.; Davenport, M.L.; Canedy, C.L.; Vurgaftman, I.; Kim, M.; Kim, C.S.; Merritt, C.D.; Bewley, W.W. Interband cascade laser on silicon. *Optica* **2018**, *5*, 996–1005. [CrossRef]
25. Yang, C.-A.; Xie, S.-W.; Zhang, Y.; Shang, J.-M.; Huang, S.-S.; Yuan, Y.; Shao, F.-H.; Zhang, Y.; Xu, Y.-Q.; Niu, Z.-C. High-power, high-spectral-purity GaSb-based laterally coupled distributed feedback lasers with metal gratings emitting at 2 μm. *Appl. Phys. Lett.* **2019**, *114*, 021102. [CrossRef]
26. Ning, C.; Yu, T.; Liu, S.; Zhang, J.; Wang, L.; Liu, J.; Zhuo, N.; Zhai, S.; Li, Y.; Liu, F. Interband cascade lasers with short electron injector. *Chin. Opt. Lett.* **2022**, *20*, 022501. [CrossRef]

27. Nguyen-Van, H.; Baranov, A.N.; Loghmari, Z.; Cerutti, L.; Rodriguez, J.-B.; Tournet, J.; Narcy, G.; Boissier, G.; Patriarche, G.; Bahriz, M. Quantum cascade lasers grown on silicon. *Sci. Rep.* **2018**, *8*, 7206. [CrossRef]
28. Calvo, M.R.; Bartolomé, L.M.; Bahriz, M.; Boissier, G.; Cerutti, L.; Rodriguez, J.-B.; Tournié, E. Mid-infrared laser diodes epitaxially grown on on-axis (001) silicon. *Optica* **2020**, *7*, 263–266. [CrossRef]
29. Tournié, E.; Bartolome, L.M.; Calvo, M.R.; Loghmari, Z.; Díaz-Thomas, A.D.; Teissier, R.; Baranov, A.N.; Cerutti, L.; Rodriguez, J.B. Mid-infrared III-V semiconductor lasers epitaxially grown on Si substrates. *Light Sci. Appl.* **2022**, *11*, 13. [CrossRef] [PubMed]

photonics

MDPI

Review

Advances in Femtosecond Coherent Anti-Stokes Raman Scattering for Thermometry

Kaiyuan Song [1], Mingze Xia [2], Sheng Yun [1], Yuan Zhang [1,3], Sheng Zhang [3], Hui Ge [1], Yanyan Deng [1], Meng Liu [1], Wei Wang [1], Longfei Zhao [1], Yulei Wang [1], Zhiwei Lv [1] and Yuanqin Xia [1,3,*]

1 Center for Advanced Laser Technology, Hebei Key Laboratory of Advanced Laser Technology and Equipment, Hebei University of Technology, Tianjin 300401, China; 202221901046@stu.hebut.edu.cn (K.S.); 202121901001@stu.hebut.edu.cn (S.Y.); 18b921025@stu.hit.edu.cn (Y.Z.); 202311901004@stu.hebut.edu.cn (H.G.); 202111901002@stu.hebut.edu.cn (Y.D.); 202221102006@stu.hebut.edu.cn (M.L.); 202321901047@stu.hebut.edu.cn (W.W.); 202321103006@stu.hebut.edu.cn (L.Z.); wyl@hebut.edu.cn (Y.W.); zhiweilv@hebut.edu.cn (Z.L.)

2 CNOOC Energy Technology and Services, Tianjin 300450, China; xiamz@cnooc.com.cn

3 National Key Laboratory of Science and Technology on Tunable Laser, School of Astronautics, Harbin Institute of Technology, Harbin 150001, China; sh_zhang@hit.edu.cn

* Correspondence: xiayq@hebut.edu.cn

Abstract: The combustion process is complex and harsh, and the supersonic combustion flow field is also characterized by short duration and supersonic speed, which makes the real-time diagnostic technology for the transient environment extremely demanding. It is of great significance to realize high time-resolved accurate measurement of temperature, component concentration, and other parametric information of the combustion field to study the transient chemical reaction dynamics of the combustion field. Femtosecond CARS spectroscopy can effectively avoid the collision effect between particles in the measurement process and reduce the influence of the non-resonant background to improve the measurement accuracy and realize the time-resolved measurement on a millisecond scale. This paper introduces the development history of femtosecond CARS spectroscopy, points out its advantages and disadvantages, and looks forward to the future development trend to carry out high time-resolved measurements, establish a database of temperature changes in various complex combustion fields, and provide support for the study of engine mechanisms.

Keywords: laser spectroscopy; coherent anti-Stokes Raman scattering; femtosecond single-pulse CARS; femtosecond time-resolved CARS spectroscopy; temperature measurement

1. Introduction

Laser spectroscopy is the main non-contact method in high-temperature combustion temperature measurement [1,2]. The most significant advantage of this approach is that it neither interferes with the combustion field being measured nor is it affected by the combustion flow field [3,4]. Currently, high-temperature turbulent flame temperature measurements are conducted using infrared laser-absorption spectroscopy sensors [5,6], which provide temperature measurements with response times on the order of microseconds to milliseconds and a temperature measurement range of 400 K–3000 K. However, the limited spatial resolution and low temporal resolution of this method have led to an increase in the uncertainty in the high-temperature turbulent flame temperature measurements [7]. In addition, laser-induced fluorescence [8–10], laser-induced thermal grating spectroscopy [11], and tunable diode laser absorption spectroscopy [12] can also be used, but these methods also have limitations.

The femtosecond coherent anti-Stokes Raman scattering (CARS) spectroscopic technique has received much attention due to its non-interference characteristics, high accuracy [13], and high time resolution [14–16]. Although the CARS spectroscopic combustion field temperature measurement technique based on nanosecond laser sources is well established and widely used [17–19], it is nevertheless subject to the zero point non-resonant signal and collision effect due to the time scale of nanosecond, which in turn reduces the measurement accuracy and sensitivity of CARS spectroscopy technology [20]. Furthermore, the repetition frequency of nanosecond lasers is typically in the range of 10–20 Hz, which allows for only a few tens of temperature measurements per second. This limitation makes it challenging to apply the technique to transient temperature measurements in high-temperature and high-pressure turbulent combustion fields.

With the development of laser technology, ultrashort pulsed lasers have shown great potential in scientific research and industrial applications. Compared with traditional lasers, ultrashort pulsed lasers have pulse widths of femtoseconds or even attoseconds [21], peak powers of more than GW, and good temporal and spatial coherence. These characteristics make ultrashort pulsed lasers show unique advantages and broad prospects in many fields. In material processing, the ultra-high peak power and transient heating effect of ultrashort pulsed lasers can be used for precision micro–nano-manufacturing, and new component preparation can also be realized through cold material processing [22]. In the biomedical field, the low thermal effect and high spatial resolution of ultrashort pulsed lasers make them an effective tool for non-destructive biphotonic imaging and precision surgery [23]. In addition, ultrashort pulsed lasers show unique technical advantages and application prospects in high-precision probing, precision grating inscription, and carrier dynamics [24,25]. Especially, in combustion temperature measurement, the combination of CARS with femtosecond lasers enables the acquisition of thousands of temperature measurements per second, thereby making it possible to carry out real-time temperature measurements of transient combustion processes, such as the working cycle of internal combustion engines and jet flames. In addition, some scholars have also studied the nanosecond and femtosecond hybrid CARS techniques and have achieved some results [26–28].

Approximately 90 percent of the world's energy supply today is generated by combustion. In the combustion field, the temperature of the flame affects the pathway and concentration of the chain reaction of various components in the field, and obtaining the temperature information of the combustion field can provide a significant basis for improving the efficiency of fuel combustion and the design of combustion devices. As a technology that can achieve transient temperature diagnosis in complex combustion scenarios, femtosecond CARS technology is necessary to summarize and prospect the research progress of this technology. In this paper, the research progress on applying femtosecond CARS temperature measurement technology in various combustion fields is reviewed and expected to provide a reference for the measurement method of high-temperature turbulent flame temperature.

2. Ultrashort Pulse CARS Spectroscopy Temperature Measurement

Femtosecond CARS spectroscopic technology is a non-contact method for measuring temperature based on the rotation spectrum of gas molecules without labeling or staining. In 1965, Make and Terhune [29] used a ruby laser to emit light, which produced an excited Raman signal as Stokes light in a benzene material. Together with the ruby laser as the pump light, this converged in the sample under test with a specific phase match, and they discovered a new four-wave mixed frequency signal, which they named CARS, and it was first used for gas-phase concentration measurements by Regnier [30] in 1974. In 1975, Moya [31] made the first temperature measurement using CARS signals of H_2 molecules. CARS spectra reflect the distribution of particles in a gas at various energy levels, and information such as the temperature, concentration, and pressure of the gas can be obtained using the spectra. Furthermore, femtosecond CARS [32] typically implies that the laser pulse itself is shorter than the characteristic collision time scale of the objectives of the test. Thereby, measurements are hardly affected by collisional-quenching effects, and femtosecond CARS spectroscopy enables the direct measurement of turbulent combustion flames with time scales between 1 and 100 μs [33,34]. Ultrafast CARS was first used for gas-phase temperature measurements by Lang et al. in 1999 [35].

Figure 1 illustrates the energy level transition in the CARS generation process. Pump light and Stokes light reach the probe sample simultaneously, exciting the particle from the ground state to the virtual state due to the pump light. Excited Raman scattering occurs under the influence of Stokes light, causing the particle to transition from the virtual state to the first excited state. In the presence of probe light, the particle transitions from the first excited state to a higher virtual state and then returns to the ground state, producing a CARS signal. According to the different detection targets, the appropriate frequency, the excitation of the corresponding Raman vibration, the CARS signal light, and the three beams of incident light frequency are used to meet the following relationship: $\omega_4 = \omega_1 - \omega_2 + \omega_3$.

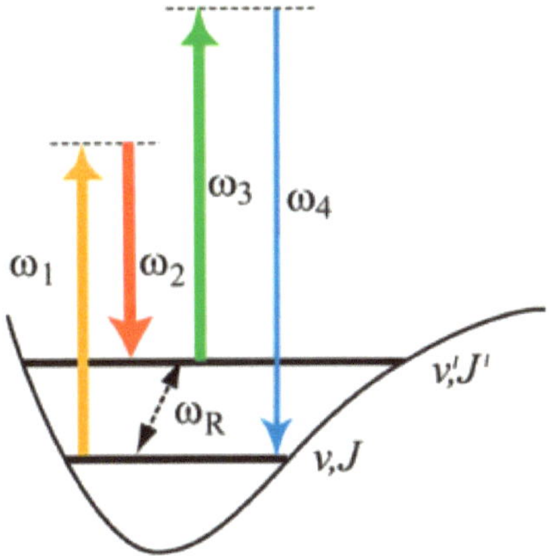

Figure 1. Schematic of molecular states and the CARS signal generation process. v,J and v',J' are different rovibrational states, and ω_R is the Raman resonance frequency of the medium molecule [32].

In addition to the aforementioned energy conservation, momentum conservation is also satisfied, expressed as $k_4 = k_1 - k_2 + k_3$, where k_1, k_2, k_3, and k_4 are the wave vectors corresponding to the pump light, Stokes light, probe light, and CARS signals, respectively. In 1978, Eckbreth [36] proposed the well-known BOX-CARS phase matching technique with Shirley et al. Subsequently, Shirley et al. [37–39] enhanced the BOX-CARS phase matching technique and proposed the folded BOX-CARS technique, as illustrated in Figure 2. This method of phase matching allows the CARS signal to be separated from the direction of the incident light in three-dimensional space, effectively preventing the incident light from interfering with the CARS signal measurement and further improving the spatial resolution of the CARS technique.

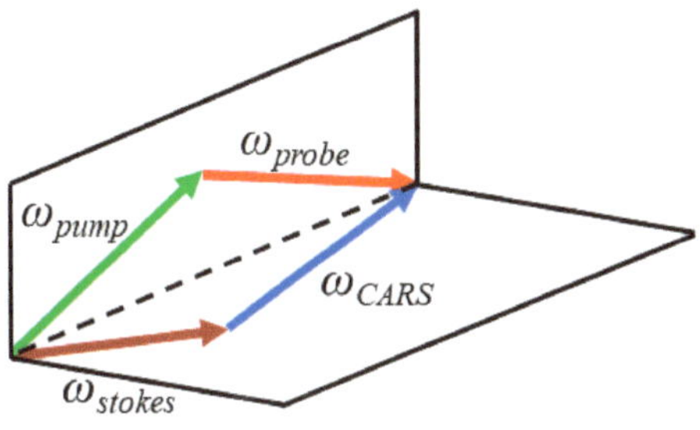

Figure 2. Folded BOX-CARS phase matching method.

In recent years, with the rapid development of femtosecond laser technology and the maturity of high-power femtosecond laser products, people have gradually paid more and more attention to the study of femtosecond CARS spectroscopic temperature measurement technology [40–46], which has been applied in complex combustion environments, such as restricted spaces [47], turbulent flames [3], and other scenarios. After obtaining CARS spectra, the time and frequency domain expressions of the CARS signal intensity and the frequency domain electric field model can be simulated using MATLAB R2018a software, and the characteristics of the CPP-fs-CARS spectrum can be simulated and analyzed by debugging the specific parameter values in the equations to adjust the degree of chirp, resonance/non-resonance ratio, spot size, etc., and to build up the best theoretical model at room temperature. After determining the specific values of other parameters, it is only necessary to change to different temperatures in the theoretical spectra to build a library of theoretical models of CARS spectra at various temperatures. The experimentally collected spectrograms at high temperatures are compared with the established optimal theoretical spectrograms at different temperatures to extract the high-temperature flame temperature. The femtosecond CARS spectroscopic thermometry technique includes three variants of the techniques in experiments, namely, time-resolved femtosecond CARS spectroscopic thermometry, chirped probe pulse femtosecond CARS spectroscopic thermometry, and hybrid femtosecond/picosecond CARS spectroscopic thermometry. These are described in the following sections.

2.1. Femtosecond Time-Resolved CARS Spectroscopy

In CARS spectroscopy, in addition to the CARS signal, various background noises, such as scattered light, autofluorescence of the sample, and non-resonant signals, are also detected. Among them, the non-resonant background noise comes from the non-resonant part of the third-order nonlinear polarization rate [48]. Since the wavelength of the non-resonant background noise is very close to that of the CARS signal, it is difficult to eliminate it by filtering methods, and these noises affect the spectral resolution and detection sensitivity of CARS spectra. Therefore, several methods have been proposed to suppress the non-resonant background noise, such as the polarization-sensitive detection method [49], the back detection method [50], the phase shaping method [51], and the time-resolved method [52–54]. The time-resolved CARS (T-CARS) method can eliminate the non-resonant background noise by taking advantage of the different dephasing times of the resonant signal and the non-resonant background noise [55].

Femtosecond CARS is mainly based on the decay of Raman coherence after initial excitation of the pump and Stokes pulses to temperature measurements [56–60]. The reason for the Raman coherence decay is that the individual Raman transitions are slightly different in oscillation frequency and no longer interfere constructively [61,62]. Because of the number of Raman transitions with significant population increases with temperature, temperature measurements can be carried out this way. Time-resolved CARS spectroscopy is based on the conventional CARS spectroscopic temperature measurement technique, with the addition of a high-precision linear stage, using the frequency difference between the pump pulse and the Stokes pulse to stimulate the Raman mode of the sample and detecting it with the probe light. By continuously varying the time the probe light reaches the sample to be measured, it is possible to obtain a series of CARS spectra on the femtosecond time scale. This allows the time-resolved CARS signal to be recorded, which provides insight into the trend of signal intensity over time, and this time-dependent signal strength can be fitted by building a theoretical model to obtain the temperature information of the target sample. A typical experimental setup is shown in Figure 3, and Table 1 demonstrates the research progress of time-resolved CARS thermometry in recent years.

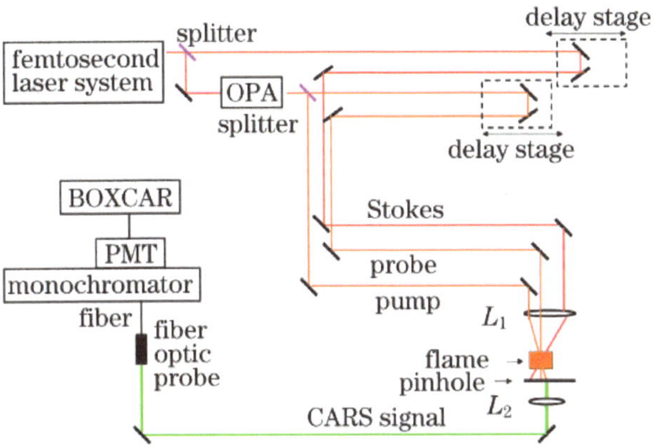

Figure 3. Schematic diagram of the time-resolved femtosecond CARS experimental setup [63].

Table 1. Research progress of time-resolved CARS temperature measurement technology.

Researcher	Year	Target Molecules	Application	Temperature/K	Accuracy	Precision
Motzkus [35]	1999	H_2	Sealed-off quartz cell	300–1100	/	/
Lucht [59]	2008	H_2	Hencken burner	1500–2500	1.6–2.7%	2–3.3%
Lucht [64]	2009	H_2	Hencken burner	300–2400	1–6%	1.5–3%
Xia [63]	2012	N_2	Methane/O_2/N_2 flame	300–1325	/	/
Song [65]	2024	N_2	Swirl burner	850–2000	/	~3.7%

The idea of using femtosecond time-resolved CARS to measure combustion field temperatures was first proposed and demonstrated to be feasible in 1999 by Motzkus et al. [35] at the Max-Planck-Gesellschaft. Then, they also achieved temperature measurements from 300 K to 1100 K by time-resolved spectroscopy of H_2 with a measurement error of 30 K [58,66]. They also measured the time-resolved spectra of signal changes at different pressures [67].

In 2006, Lucht et al. [62] at Purdue University obtained CARS signals for gas samples using probe pulses with different delay times and measured a heated gas cell in the temperature range of 300 K–940 K. In 2008, they extended the temperature range to within 1500 K to 2500 K [59] with an accuracy of $\pm$40 K and a precision of $\pm$50 K. In 2009, they used time-resolved spectroscopy followed by Fourier transformation to improve the accuracy and precision of temperature measurement from 300 K–2400 K to 1–6% and 1.5–3%, respectively [64].

In 2012, our team [63] conducted a study to measure the temperature of a methane/O_2/N_2 premixed flame at atmospheric pressure using the femtosecond CARS technique. A 40 fs laser pulse was employed to excite the rotational spectrum of N_2, with the CARS signal being measured within a few picoseconds of the initial coherent excitation. The flame temperatures were measured at 300 K–1325 K. The results were in good agreement with theoretical calculations and exhibited good repeatability.

However, femtosecond time-resolved CARS has some inherent drawbacks. Firstly, the traditional time-resolved CARS spectroscopy system is complex and expensive [68,69], and secondly, the measurement process needs to take points several times, which makes the technique unable to achieve the measurement of transient processes. Furthermore, for samples containing multiple components to be measured, the center wavelength of the laser needs to be changed continuously to obtain the CARS spectral information of each component in the sample, which is not conducive to practical applications [70–72]. Due to these limitations, this technique has not been studied much in temperature measurement, while femtosecond single-pulse CARS thermometry has received attention, which mainly includes femtosecond/picosecond hybrid CARS spectroscopy and chirped probe pulse femtosecond CARS spectroscopy, which allows for transient measurements of the combustion field with a high time resolution.

It is worth noting that there are also scholars who are looking for some solutions to the above shortcomings. In 2024, Song [65] introduced additional probe pulses into the conventional time-resolved CARS system [52], and the dual-probe scheme allows for obtaining more spectral information in a single measurement [9,73]. In this dual-probe scheme, the pump pulse and the Stokes pulse interact with the N_2 molecule simultaneously, while the two probe pulses are positively delayed by 3.44 ps and 2.88 ps, respectively, and the temperature of the system can be determined by comparing the intensity ratios of the different CARS signals generated. Thus, the stepwise scan on the relaxation of the vibrational coherence is no longer required when performing temperature measurements. Two probe beams of different wavelengths will produce two CARS signals of different wavelengths that can be distinguished in the same spectrum so that single-shot measurement can be realized in principle [66]. However, at present, the method only has good precision and accuracy in monitoring flame temperatures below 2000 K. This is mainly due to the large fluctuations in CARS intensity above 2000 K, which can directly affect the measurement accuracy. Overall, this method is a potential candidate for accurate monitoring of real-time temperatures in turbulent combustion.

2.2. *Femtosecond Single-Pulse CARS Technology*

Typical femtosecond laser spectra have a spectral width of about a few tens of nanometers (about 200 cm^{-1}). When the incident pump light and Stokes light are both broad spectrum femtosecond pulses, some frequencies in the pump pulse may meet the same frequency difference with several frequencies in the Stokes pulse, and multiple pairs of Raman transitions at the same frequency can be excited at the same time. At the same time, the fixed-frequency component of the pump light may also meet the requirement of coherent excitation with many different frequencies in the broad spectrum of Stokes and can scan Stokes light frequencies without adjusting the time delay to realize the coherent excitation of Raman transitions of molecules at different vibrational/rotational energy levels, which is the femtosecond single-pulse CARS thermometry technology. Femtosecond single-pulse CARS techniques can be classified into two categories: chirped probe pulse CARS, which uses a short chirped pulse as the excitation source, and hybrid femtosecond/picosecond CARS, which uses femtosecond main pulses and picosecond pulses. These two schemes enable high time resolution temperature measurements and overcome the technical constraints of conventional CARS systems. They are powerful tools for studying combustion dynamics.

2.2.1. Femtosecond Chirped Probe Pulse CARS

Femtosecond chirped probe pulse CARS (CPP-fs-CARS) is a novel single-pulse CARS spectrometry technology. Based on the common femtosecond CARS, CPP-fs-CARS utilizes two synchronously locked femtosecond lasers. One laser generates the main pulse with stable intensity, while the other generates a short chirped probe pulse. The chirped pulses are synchronized in time with the main pulse, and the time domain stretching and frequency domain chirping of the probe pulse are achieved by adding dispersive media to the probe optical path, which stretches the probe pulse width from femtoseconds to picoseconds while the different frequency components are separated on the time axis. The Raman coherent wave packet, driven by a single femtosecond pump pulse, interacts with different frequency components of the probe pulse at different times, resulting in the corresponding CARS response in the frequency domain. Unlike conventional time-resolved CARS, which uses delay line scanning, CPP-fs-CARS uses a single pulse to directly acquire time series data, greatly simplifying the system. Figure 4 shows a typical experimental setup.

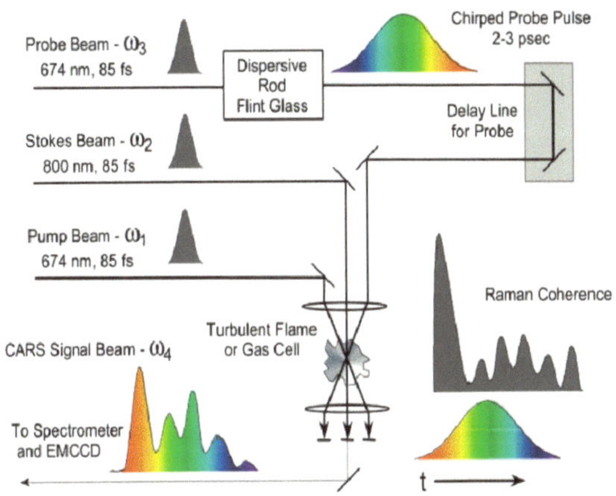

Figure 4. Schematic diagram of the single-laser-shot CPP-fs-CARS measurements [74].

The CPP-fs-CARS technique was initially proposed in 2002 by Lang and Motzkus et al. at the Max-Planck-Gesellschaft [75]. In their experiments, they introduced chirp using 6 cm SF-10 glass rods, which broadened the 100 fs probe pulse to 500 fs. The CARS signal also had a chirp effect, enabling the mapping of the time curve to the frequency domain. The results indicate that the transient shape agrees well with multiple measurements in the region where the delay time exceeds the duration of the modulated pulse, with a small phase shift. By applying this technique to a single temperature measurement of H_2, a temperature measurement range of 300 K to 1100 K was achieved with an error of 30 K, which is consistent with the theoretical fit. Table 2 demonstrates the research progress of CPP-fs-CARS thermometry in recent years.

Table 2. Research progress of CPP-fs-CARS temperature measurement technology.

Researcher	Year	Target Molecules	Application	Temperature/K	Accuracy	Precision
Lang [75]	2002	H_2	Combustion cells	300–1100	~2.7%	/
Roy [64]	2009	N_2	Near-adiabatic flame	300–2400	1–6%	1.5–3%
Lucht [76]	2011	N_2	Hencken burner	1790–1940	~2%	5%
Lucht [77]	2016	N_2	Dual-swirl gas turbine model combustor	300–2200	±3%	±2%
Lucht [78]	2017	N_2	Hencken burner	295–2295	2.7%	±3.5%
Thomas [79]	2019	N_2	Turbulent spray flames	2512	2.8%	±3.4%
Lucht [13]	2021	CO_2/N_2	Hencken burner	295–1420	N_2: 1.1–8.9% CO_2: 0.6–5.3%	1.6% (>1200 K) 1.1–1.4% (<1200 K)
Chang [14]	2023	H_2	High-pressure rocket chamber	2000–3000	/	/

Roy and Lucht at Purdue University have carried out extensive thermometry work using CPP-fs-CARS. In 2009, they determined the temperature from fitting a single spectrum of the CARS signal by varying the delay time of the probe pulse for the pumping and Stokes excitation (~2 ps) [64]. The accuracy and precision of the method were 1–6% and 1.5–3% in the temperature range of 300 K–2400 K, respectively. A single temperature measurement of gas-phase N_2 was achieved for the first time at a rate of 1 kHz by fitting the experimental results to the theoretical calculations by the least-squares method. Compared to H_2, N_2 is more suitable for temperature measurement of combustion gas streams. This work thus extends the application of the technique with the promise of application to transient and turbulent systems but has not yet applied it to actual combustion fluids and has only validated it under ideal conditions.

To validate the technique in a practical situation, in 2011, Lucht and Richardson [76] performed 1000 Hz temperature measurements on a flame driven by a piston and a turbulent methane air flame. A theoretical model of CPP-fs-CARS considering the parameters of the pump pulse, the Stokes pulse, and the probe pulse led to the development of resonant and non-resonant polarization contributions. The standard deviation of the technique in gas cells and laminar flames is less than 2% of the mean temperature. Periodic fluctuations of 10 Hz were accurately captured in piston-driven flames, validating the ability of the technique to detect transient phenomena. Then, they discussed suppressing the non-resonant background by polarization techniques for CPP-fs-CARS measurements [80], providing a methodology for the further development of high-fidelity CPP-fs-CARS models.

The vibrational modes of methane are easier to detect because the methane scattering cross-section is eight times larger than the scattering cross-section of N_2 [81]. In 2014, Dennis modified the objective function of the spectral fitting algorithm to calculate the energy levels of N_2 on top of Richardson's work so that the temperatures of N_2 and methane could be fitted simultaneously. Subsequently, Dennis used the modified code to perform 5 kHz femtosecond CARS temperature measurements in an out-of-core jet diffusion flame [82], a swirl-stabilized combustor [83], and a gas turbine model combustor [84], respectively.

Another validation was carried out in 2016 by Dennis and Lucht [77] in a more practically relevant high-temperature turbulent flame. Measurements were made at 73 points within the gas turbine model combustion in different steady-state environments, and a CARS signal sufficient for analysis was obtained for almost every laser hit. The spatial resolution of the single-laser temperature measurements was about 600 μm, the precision was about ±2%, the accuracy was about ±3%, and the dynamic range was sufficient for temperature measurements from 300 K to 2200 K. Lucht subsequently improved the spectral fitting method based on the statistical method of maximum likelihood to achieve a temperature accuracy and precision of 2.7% and ±3.5% for flame temperature and 9.9% and ±6.1% at room temperature [78] and devised a scientific method to evaluate the system's performance. Results from multiple sets of laser parameters are combined to generate an error-weighted temperature from the top-performing calibrations. This method provides a basis for the further application of the technique to actual spray combustion processes.

Thomas at Purdue University and Lowe at the University of Sydney also measured temperature in turbulent spray flames using the CPP-fs-CARS technique with a repetition rate of 5 kHz. In 2019, they performed temperature measurements in spray turbulent flames with ethanol droplets of varying concentrations, achieving an integration time of 3 ps and a spatial resolution of ~800 μm along the direction of beam propagation and lateral dimensions of ~60 μm of the spatial resolution [79]. The accuracy of the technique was verified with a relative accuracy and precision of 2.8% and ±3.4%, respectively, at peak flame temperatures of up to 2512 K. The results show the accuracy of the technique. The spray flame data were processed to obtain the average axial and radial temperature distributions in both the dilute and dense spray flames. The observed temperature range was between 1000 K and 2000 K, as shown in Figure 5.

They then went on to measure the temperature fields of ethanol and acetone spray flames separately using this technique [85]. Flames with different spray densities were obtained by varying the nozzle recession length (0 mm–80 mm), as shown in Figure 6, where the nozzle recession length significantly affects the flame structure at $x/D > 10$. Temperature probability density function analysis showed that the dense spray inhibited the inhalation of hot flying bodies into the spray envelope, reducing the vaporization and combustion speed rates. With a temperature measurement range of about 300 K–2500 K and a temperature measurement accuracy of ±2%, this method was used to measure, for the first time, the transient temperature fields of ethanol and acetone spray flames under different spray density conditions using a high-speed CPP-fs-CARS.

With the increase in carbon-containing fuels, researchers gradually began CARS temperature measurements using carbon dioxide (CO_2) as a probe molecule. Lucht developed the phenomenological model of CO_2/O_2 femtosecond CARS in 2021 [13]. Validation experiments in different CO_2/O_2 gas mixtures demonstrated that O_2 interference could be eliminated at a probe pulse delay of approximately 9 ps. Flame temperature tests were conducted on a Hencken burner, and the results showed that CO_2 femtosecond CARS can accurately measure temperatures under 1200 K, with up to 1% accuracy under 1100 K. Additionally, precise measurements can be obtained with a single set of laser parameters by increasing the probe pulse delay of the CO_2 femtosecond CARS. The CO_2 femtosecond CARS technique offers several advantages. Firstly, it is not affected by interference from O_2 and non-resonant backgrounds. Secondly, it is more sensitive to flames at lower temperatures (<1000 K).

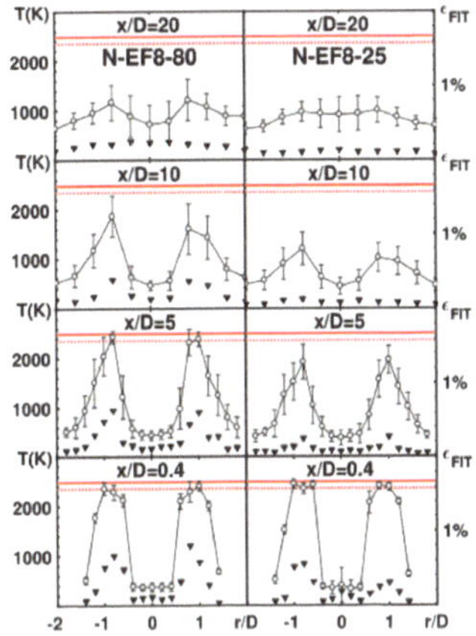

Figure 5. The radial profile of the mean temperature of the dilution spray (N-EF8-80) and the dense spray (N-EF8-25) at x/D = 0.4–20; x/D represents the length ratio to total length, the inverted triangles represent fitting errors, and circles represent temperature [79].

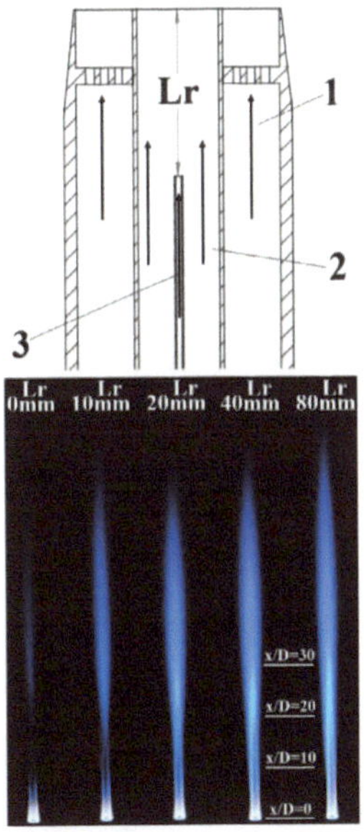

Figure 6. Long-exposure (1 s) images of samples with acetone flame at different recess lengths; x/D represents the length ratio to total length, with the pilot stream in channel 1, the main air jet stream in channel 2, and the liquid fuel stream in channel 3 [85].

Lucht and Chang [14] recently demonstrated hydrogen CPP-fs-CARS temperature measurements in a high-pressure model rocket motor combustion chamber. For the

Hencken burner flame, time-averaged spectra were taken at 144 Hz for an improved SNR. Around 2000 spectra were averaged and fit using the fitting algorithm. Figure 7 below shows an excellent fit between the experimental and theoretical spectra. They also acquired CPP-fs-CARS spectra along the axial direction of the combustion chamber under gaseous H_2/O_2 fuel conditions at pressures of up to 7 MPa and an equivalence ratio of 3.0. This was the first time that the CPP-fs-CARS technique was applied to high-pressure rocket motor experiments that yielded transient temperature field information with a high time resolution, demonstrating the feasibility of using this technique to measure high-pressure rocket motor temperatures.

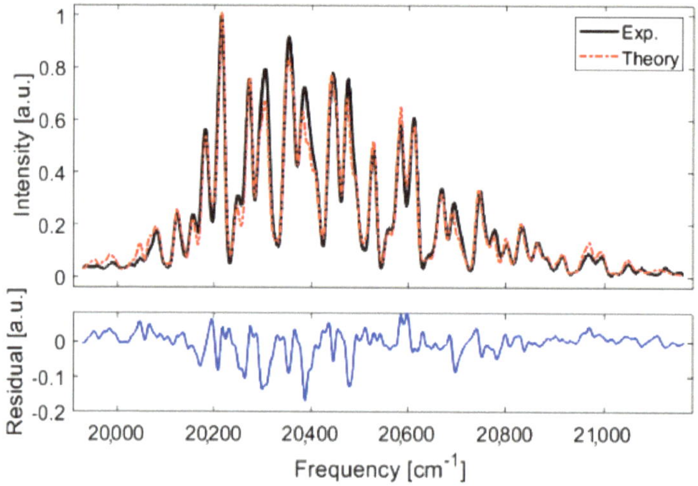

Figure 7. H_2 CPP-fs-CARS spectra in a Hencken burner flame with theoretical fit at an adiabatic flame temperature of 2129 K and 0.1 MPa (1 atm), where the blue lines represent the residuals, meaning the difference between the actual observations and the model predictions [14].

The chirped pulses in the CPP-fs-CARS technique are obtained based on the dispersion effect extension, and the spectrum of the signal produced by this method has the inherent limitation of a low resolution. In addition, their complex spectra increase the difficulty of the fitting process. To overcome this limitation, the researchers proposed a hybrid femtosecond/picosecond CARS technique using a pulse shaper to generate narrow linewidth picosecond probe pulses. By independently controlling the time domain and frequency domain parameters of the probe pulse, this method realizes the organic combination of the high temporal resolution and high spectral resolution, which compensates for the inherent limitations of CPP-fs-CARS and further extends the application range of the single-pulse CARS technique.

2.2.2. Hybrid Femtosecond/Picosecond CARS Spectroscopy for Temperature Measurement

The hybrid femtosecond/picosecond (fs/ps) CARS technique employs a pulse shaper to generate picosecond probe pulses with narrower linewidths in the frequency domain, which have a significantly higher spectral resolution than chirped pulses obtained using dispersion effect expansion [86,87]. As a result, the hybrid fs/ps CARS system can measure purely rotationally dynamic CARS processes, whereas the broadband CPP-fs-CARS system struggles to achieve the same level of high-precision rotational spectral detection. In addition, each peak in the hybrid fs/ps CARS spectra directly corresponds to a specific rotational Raman jump, which more intuitively reflects the rotational temperature information. Compared with CPP-fs-CARS, which requires complex iterations to fit the entire spectral curve, the hybrid fs/ps CARS spectra are simpler to fit and less computationally intensive [88]. It can be seen that the high spectrally resolved probe pulses generated by pulse shaping are the keys to achieving high temporal and spectral resolutions in hybrid fs/ps CARS. This novel single-pulse excitation strategy refines the advantages of the femtosecond CARS spectroscopy technique in terms of accuracy and ease of use, and a typical experimental setup is shown in Figure 8.

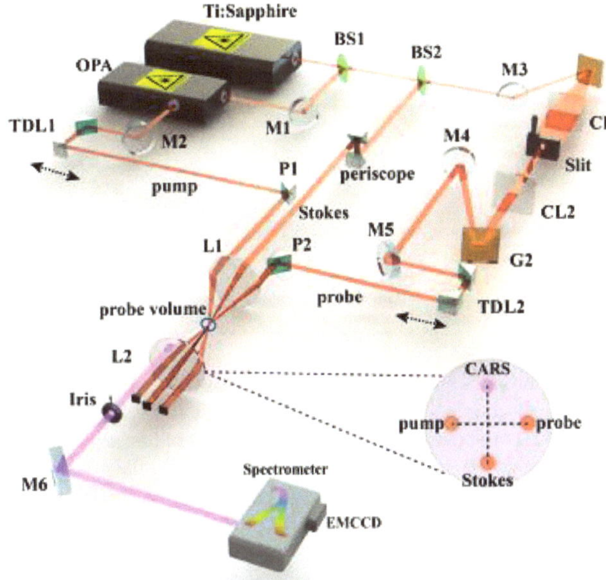

Figure 8. Hybrid femtosecond/picosecond CARS experimental setup [89]. OPA: optic parametric amplifier; BS1-2: beam splitter; L1-2: lens; M16: mirror; TDL1-2: tunable delay line; G1-2: diffraction grating; P1-2: prism; CL1-2: compensation lens; EMCCD: electron-multiplying charged coupled device.

Hybrid fs/ps CARS technology is also divided into two main categories, namely, purely vibrational hybrid fs/ps CARS and purely rotational hybrid fs/ps CARS. Purely vibrational hybrid fs/ps CARS technology uses femtosecond pump pulses and picosecond probe pulses of similar center wavelengths to excite vibrational excitations of the gas, generating the corresponding vibrational CARS signals. Unlike conventional CARS systems that scan the laser wavelength, a full spectrum of vibrational radiation can be obtained by independently controlling the parameters of the femtosecond and picosecond laser pulses. In particular, adjusting the time delay of the picosecond probe pulse controls the temporal resolution of the CARS process, and varying the relative intensity of the picosecond pulse controls the suppression of the non-resonant background. Vibrational CARS spectral thermometry shows good accuracy in the temperature range above 1000 K. Pure rotation hybrid fs/ps CARS, on the other hand, uses femtosecond and picosecond pulses with large wavelength differences to excite different rotational transitions, and this technique can distinguish the transitions between different rotational quantum numbers and provide high-resolution rotational Raman spectra, thus accurately measuring the rotational temperature of the gas, which shows high sensitivity at lower temperatures.

Hybrid fs/ps CARS spectroscopy was proposed by Prince and Chakraborty et al. [90] in 2006. Extensive research in hybrid fs/ps CARS temperature measurement has been performed by Miller and Meyer. In 2010, they performed high-speed temperature measurements in a high-temperature gas-phase system [91]. Using a 100 fs broadband pump and Stokes light to excite the vibrational energy levels of N_2, the Raman resonance response was probed with a frequency-narrowed 2.5 ps probe pulse (delayed by 2.36 ps from the pump and Stokes pulses), which reduced the non-resonant background by two orders of magnitude. Experimentally obtained spectra were performed in steady-state and pulsating H_2–air flames at a rate of 500 Hz, exhibiting a temperature precision of 2.2 percent and an accuracy of 3.3 percent at 2400 K, enabling fast and precise measurements of high-temperature gas-phase temperatures. Table 3 demonstrates the research progress of hybrid femtosecond/picosecond CARS thermometry in recent years.

Table 3. Research progress of hybrid femtosecond/picosecond CARS temperature measurement technology.

Researcher	Year	Target Molecules	Application	Temperature/K	Accuracy	Precision
Miller [91]	2010	N_2	Unsteady high-temperature flames	2400	~3.3%	2.2%
Miller [92]	2011	N_2	The time delay is 13.5 ps to 30 ps	306–700	/	1%
Miller [73]	2014	$O_2/H_2/N_2$	Adiabatic H_2–air Hencken burner flame	298–2300	/	RCARS: 5% VCARS: 2%
Kearney [41]	2015	N_2	Near-adiabatic H_2/air flames Premixed C_2H_4/air flames	H_2: 1550 C_2H_4: 1660	/	1–1.5%
Escofet-Martin [93]	2020	N_2	Pressure (0.9–1.1 bar)	280–310	0.62%	0.42%
Li [89]	2020	N_2	Hencken burner	2110	1.2%	/
Kim [94]	2023	C_2H_4	Supersonic combustion facility	2294	/	/

Miller then used the same broadband pump and Stokes pulses to excite the molecular rotational energy levels. Raman coherence was then probed using a frequency-narrowed 8.4 ps probe pulse after a 13.5 ps time delay to eliminate non-resonant background interference [92]. The technique enables the direct measurement of the collisional out-of-phase rate for each rotational energy level (J-value) within the range of 13.5 ps–100 ps. The results demonstrate that the temperature measurement error is less than 1% at atmospheric pressure when the probe pulse delay is less than 30 ps, indicating no collisional effect. Frequency and time domain model calculations demonstrate that, under current conditions, the fs/ps CARS achieves a best-fit temperature accuracy of up to 1% within the range of 306 K–700 K at a 13.5 ps time delay, as demonstrated in Figure 9. This method allows for fast and accurate gas temperature measurements.

In 2014, Meyer [73] developed a dual-pump hybrid fs/ps CARS system capable of exciting both rotation–vibrational and pure rotation transitions of multiple molecules simultaneously. The technique utilizes two beams of 100 fs femtosecond pump pulses at 660 nm and 798 nm to simultaneously excite N_2/CH_4 vibrational and $N_2/O_2/H_2$ rotational Raman coherence. A common beam of 798 nm femtosecond Stokes pulse is used, with a narrow-band 798 nm picosecond probe pulse detecting all the Raman resonance responses after a certain time delay to suppress non-resonant background and collisional effects. The system achieves quantitative detection of four key combustion species, namely, N_2, O_2, CH_4, and H_2, as well as accurate temperature measurement over a wide temperature range from 298 K to 2300 K through broadband excitation and multiplexing of multiple molecular transitions, and the temperature measurements were precise, with an accuracy of within 2% of the theoretical predictions for equilibrium.

In 2015, Kearney [41] employed second harmonic bandwidth compression (SHBC) to produce high-energy sub-10 ps probe pulses for collision-free and non-resonant background measurements at kilohertz rates. For a peak SNR of more than 60, attained for temperatures of up to T = 1660 K in the C_2H_4/air flame of the McKenna burn-errand T = 1550 K in the near-adiabatic H_2/air flame stabilized on the Hencken burner, temperature-measurement precision is 1–1.5%, which is a factor of 3–4 better than the best nanosecond rotational CARS measurements [95,96] at atmospheric pressure. This temperature measurement precision is comparable to the best results measured using femtosecond vibrational CARS schemes at flame temperatures above 2000 K [74,92] and meets or exceeds the accuracy shown using rotating fs/ps CARS in low-temperature environments in air [97] and pure N_2 [98,99].

To predict and interpret the effect of spectrally significant modulation of the probe pulses on the hybrid fs/ps pure rotation CARS results, Yang of the Chinese Academy of Engineering Physics and Escofet-Martin at the University of Edinburgh [100] used a home-built SHBC in 2017 to generate picosecond pulses with different spectrally significant modulations. The simulation results with the standard model and modified model are shown in Figure 10. For the fitting of every probe delay, the residuals with the chirped model are smaller than those with the original model. However, the probe pulse chirp affects the sum of the absolute residuals, and it takes a certain amount of temperature bias to match the valley values of non-chirp summations of absolute residuals.

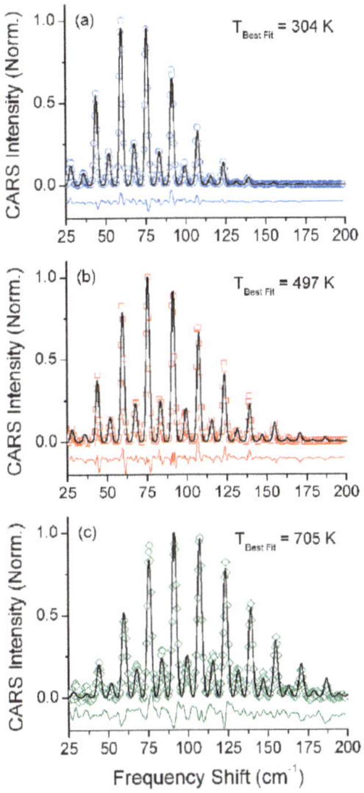

Figure 9. Single-shot pure rotational hybrid fs/ps CARS spectra of N_2 at (**a**) 306 K, (**b**) 500 K, and (**c**) 700 K, where open symbols represent experimental data while solid lines represent best-fit simulations [92].

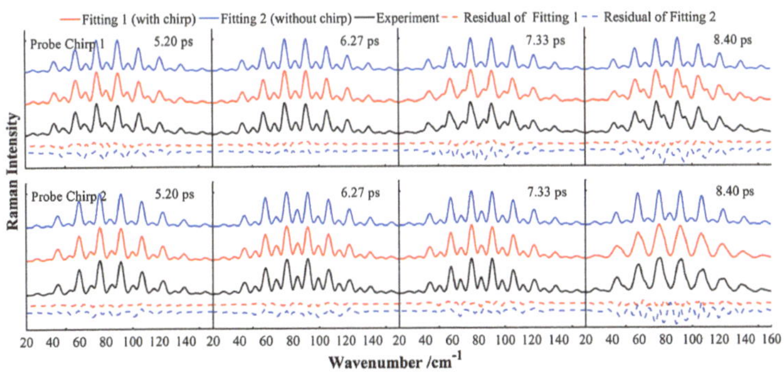

Figure 10. Comparison between the fitting results of the modified model (with probe chirp) and the original model (without chirp). Two groups of coherent anti-Stokes Raman scattering spectra of N_2 (black solid lines) with different probe chirps are fitted [100].

In 2020, Escofet-Martin [93] introduced another novel dual-detection one-dimensional fs/ps hybrid rotation CARS spectroscopy technique for simultaneous single-shot temperature, pressure, and O_2/N_2 concentration ratio measurements in the gas phase. The technique was used to extract the time and frequency domain information simultaneously and to make accurate and precise pressure measurements based on the strong correlation between the CARS signal and the pressure in the time domain and temperature measurements in the frequency domain. Highly accurate (<1%) pressure measurements in the pressure range of 1 bar–1.5 bar and the temperature range of 280 K–310 K, with spatiotemporal accuracies of 0.62% for temperature and 0.42% for pressure and the ability to characterize pressure gradient variations of 0.04 bar/mm, were achieved, demonstrating the promise of the novel technique for application in high-pressure, high-temperature, and high-concentration gradient settings.

In the same year, Li [89] at Tsinghua University proposed a hybrid fs/ps CARS thermometry method using an optimized 10 ps-35 ps probe time delay. The optimal time delay, corresponding to the maximum value of the differential at each temperature, was found by calculating the differential of the spectra at neighboring temperatures. The CARS spectrum exhibits another distinct peak when using the optimal time delay. This facilitates the fitting process and avoids temperature errors caused by a slight drift in the probe center wave number. The ability to maximize temperature sensitivity in the range of 300 K to 2500 K shows great potential in applications requiring high sensitivity. Then, they proposed a quasi-common-path SHBC method [101] to achieve spectral filtering and introduce linear dispersion through the same 4f shape filter. This method resulted in a high-energy narrow-band picosecond pulse with a center wavelength of 401.5 nm, a bandwidth of about 7 cm^{-1}, and a pulse energy of 240 μJ from a 35 fs broadband pulse. The given pulse served as a probe pulse for a hybrid fs/ps vibrating CARS system. This system allowed for single-pulse temperature measurements of methane/air flames at a rate of 1000 frames per second. The temperature uncertainty of the single-pulse measurements was less than 1%, with an accuracy of 2% at the optimal probe pulse delay time for a 1790 K flame. The method demonstrates excellent accuracy and sensitivity in harsh environments where fast temperature measurements are required.

CARS has been extensively used in harsh combustion environments, such as high-pressure gas turbines and rocket combustors [102–106]. In 2023, Dedic [94] used the fs/ps CARS technique to measure the temperature inside the University of Virginia Supersonic Combustion Facility (UVASCF) [107,108]. Previous CARS measurements in the UVASCF have used nanosecond laser excitation [109,110]. The UVASCF requires multiple viewports in the combustion chamber wall for spectral measurements using laser light. However, the interaction of the femtosecond laser with these glass windows produces a supercontinuum spectrum that affects the measurement accuracy. It reduced the generation of supercontinuum spectra by controlling the energy density of the femtosecond laser on the windows. Additionally, the optical path was optimized to improve the laser's focusing quality at the measurement point, resulting in a CARS signal of sufficient intensity. The CARS system's performance was tested in an H_2/air flame, and the results indicate that the simulated spectra are in good agreement with the single experiment spectra at 2300 K. This provides a foundation for subsequent studies of the combustion state of a dual-mode ram engine in this high-speed airflow test bed.

3. Conclusions

Femtosecond CARS spectroscopy techniques play an increasingly important role in non-contact optical temperature measurement. These techniques offer ultra-high temporal resolution for studying transient processes and non-equilibrium temperature measurements. Femtosecond CARS spectroscopy is also useful for Raman spectroscopy measurements at low wave numbers, providing information on rotational and low-frequency vibrational modes. Future developments aim to enhance spatial and temporal resolution, improve the precision and accuracy of temperature measurements, and achieve three-dimensional scanning imaging. Progress is expected in probe development, expanding measurement range, and system integration. Collaboration between researchers and industries is crucial for establishing a non-contact temperature measurement platform, which will provide more accurate technical support for complex high-temperature turbulent combustion field temperature measurements.

Author Contributions: K.S.: conceptualization, investigation, and original draft; M.X.: formal analysis; S.Y.: supervision and editing; Y.Z.: supervision and editing; S.Z.: editing; H.G.: project administration; Y.D.: project administration; M.L.: project administration; W.W.: project administration; L.Z.: project administration; Y.W.: project administration; Z.L.: project administration; Y.X.: supervision and writing—review and editing. All authors have read and agreed to the published version of the manuscript.

Funding: This research was funded by the Natural Science Foundation of China (62375074, 62075056); the Natural Science Foundation of Hebei Province in China (F2023202082, F2022202035); and the Interdisciplinary postgraduate Training Program of Hebei University of Technology (HEBUT-Y-XKJC-2022120).

Institutional Review Board Statement: Not applicable.

Informed Consent Statement: Not applicable.

Data Availability Statement: No new data were created or analyzed in this study. Data sharing is not applicable to this article.

Conflicts of Interest: Author M.X. is employed by the company CNOOC Energy Technology and Services that this research was conducted in the absence of any commercial or financial relationships that could be construed as a potential conflict of interest. The other authors declare no conflicts of interest.

References

1. Gabet, K.N.; Patton, R.A.; Jiang, N.; Lempert, W.R.; Sutton, J.A. High-speed CH_2O PLIF imaging in turbulent flames using a pulse-burst laser system. *Appl. Phys. B* **2012**, *106*, 569–575. [CrossRef]
2. Tian, Z.; Zhao, H.; Wei, H.; Tan, Y.; Li, Y. Improved opposition-based self-adaptive differential evolution algorithm for vibrational hybrid femtosecond/picosecond coherent anti-Stokes Raman scattering thermometry. *Appl. Opt.* **2022**, *61*, 4500–4508. [CrossRef] [PubMed]
3. Zheng, S.; Cai, W.; Liu, B.; Zhu, S.; Zhou, B.; Sui, R.; Lu, Q. Experimental detection of two-dimensional temperature distribution in Rocket-Based Combined Cycle combustion chamber using multispectral imaging processing. *Fuel* **2023**, *333*, 126391. [CrossRef]
4. Zheng, S.; Cai, W.; Sui, R.; Luo, Z.; Lu, Q. In-situ measurements of temperature and emissivity during MSW combustion using spectral analysis and multispectral imaging processing. *Fuel* **2022**, *323*, 124328. [CrossRef]
5. Goldenstein, C.S.; Spearrin, R.M.; Jeffries, J.B.; Hanson, R.K. Infrared laser-absorption sensing for combustion gases. *Prog. Energ. Combust.* **2017**, *60*, 132–176. [CrossRef]
6. Nair, A.P.; Lee, D.D.; Pineda, D.I.; Kriesel, J.; Hargus, W.A.; Bennewitz, J.W.; Danczyk, S.A.; Spearrin, R.M. MHz laser absorption spectroscopy via diplexed RF modulation for pressure, temperature, and species in rotating detonation rocket flows. *Appl. Phys. B* **2020**, *126*, 138. [CrossRef]
7. Braun, J.; Saracoglu, B.H.; Paniagua, G. Unsteady Performance of Rotating Detonation Engines with Different Exhaust Nozzles. *J. Propul. Power* **2017**, *33*, 121–130. [CrossRef]
8. Grib, S.W.; Fugger, C.A.; Hsu, P.S.; Jiang, N.; Roy, S.; Schumaker, S.A. Two-dimensional temperature in a detonation channel using two-color OH planar laser-induced fluorescence thermometry. *Combust. Flame* **2021**, *228*, 259–276. [CrossRef]
9. Retter, J.E.; Koll, M.; Dedic, C.E.; Danehy, P.M.; Richardson, D.R.; Kearney, S.P. Hybrid time-frequency domain dual-probe coherent anti-Stokes Raman scattering for simultaneous temperature and pressure measurements in compressible flows via spectral fitting. *Appl. Opt.* **2023**, *62*, 50–62. [CrossRef]
10. Wang, G.; Tang, H.; Yang, C.; Magnotti, G.; Roberts, W.L.; Guiberti, T.F. Quantitative laser-induced fluorescence of NO in ammonia-hydrogen-nitrogen turbulent jet flames at elevated pressure. *Proc. Combust. Inst.* **2023**, *39*, 1465–1474. [CrossRef]
11. Luers, A.; Salhlberg, A.-L.; Hochgreb, S.; Ewart, P. Flame thermometry using laser-induced-grating spectroscopy of nitric oxide. *Appl. Phys. B* **2018**, *124*, 43. [CrossRef]
12. Yu, X.; Li, F.; Chen, L.; Ou, D.; Zeng, H. Combined tunable diode laser absorption spectroscopy and monochromatic radiation thermometry in ammonium dinitramide-based thruster. *Opt. Eng.* **2018**, *57*, 026106. [CrossRef]
13. Gu, M.; Satija, A.; Lucht, R.P. CO_2 chirped-probe-pulse femtosecond CARS thermometry. *Proc. Combust. Inst.* **2021**, *38*, 1599–1606. [CrossRef]
14. Chang, Z.; Gejji, R.; Satija, A.; Lucht, R.P. Chirped-Probe-Pulse Femtosecond CARS H_2 Thermometry in a High-Pressure Model Rocket Combustor. In Proceedings of the AIAA SciTech Forum, National Harbor, MD, USA & Online, 23–27 January 2023. [CrossRef]

15. Lu, M.; Zhang, Y.; Chen, X.; Li, Y.; Wei, H. Interpulse stimulation Fourier-transform coherent anti-Stokes Raman spectroscopy: Publisher's note. *Photonics Res.* **2023**, *11*, 609. [CrossRef]
16. Gong, L.; Zheng, W.; Ma, Y.; Huang, Z. Higher-order coherent anti-Stokes Raman scattering microscopy realizes label-free super-resolution vibrational imaging. *Nat. Photonics* **2019**, *14*, 115–122. [CrossRef]
17. Roy, S.; Meyer, T.R.; Lucht, R.P.; Belovich, V.M.; Corporan, E.; Gord, J.R. Temperature and CO_2 concentration measurements in the exhaust stream of a liquid-fueled combustor using dual-pump coherent anti-Stokes Raman scattering (CARS) spectroscopy. *Combust. Flame* **2004**, *138*, 273–284. [CrossRef]
18. Marrocco, M. Comparative analysis of Herman-Wallis factors for uses in coherent anti-Stokes Raman spectra of light molecules. *J. Raman Spectrosc.* **2009**, *40*, 741–747. [CrossRef]
19. Beyrau, F.; Brauer, A.; Seeger, T.; Leipertz, A. Gas-phase temperature measurement in the vaporizing spray of a gasoline direct-injection injector by use of pure rotational coherent anti-Stokes Raman scattering. *Opt. Lett.* **2004**, *29*, 247–249. [CrossRef]
20. Roy, S.; Meyer, T.R.; Gord, J.R. Time-resolved dynamics of resonant and nonresonant broadband picosecond coherent anti-Stokes Raman scattering signals. *Appl. Phys. Lett.* **2005**, *87*, 264103. [CrossRef]
21. Xue, B.; Tamaru, Y.; Fu, Y.; Yuan, H.; Lan, P.; Mücke, O.D.; Suda, A.; Midorikawa, K.; Takahashi, E.J. A Custom-Tailored Multi-TW Optical Electric Field for Gigawatt Soft-X-Ray Isolated Attosecond Pulses. *Ultrafast Sci.* **2021**, *2021*, 9828026. [CrossRef]
22. Raciukaitis, G. Ultra-Short Pulse Lasers for Microfabrication: A Review. *IEEE J. Sel. Top. Quant.* **2021**, *27*, 1–12. [CrossRef]
23. Tsakanova, G.; Arakelova, E.; Matevosyan, L.; Petrosyan, M.; Gasparyan, S.; Harutyunyan, K.; Babayan, N. The role of women scientists in the development of ultrashort pulsed laser technology-based biomedical research in Armenia. *Int. J. Radiat. Biol.* **2021**, *98*, 489–495. [CrossRef] [PubMed]
24. Zhang, Y.; Dong, Z.; Ge, H.; Zhao, L.A.; Xu, H.; Wang, Y.; Song, L.; Xia, Y. Influence of the Surface Modification on Carrier Kinetics and ASE of Evaporated Perovskite Film. *IEEE Photonic Tech. Lett.* **2023**, *35*, 285–288. [CrossRef]
25. Lakhotia, H.; Kim, H.Y.; Zhan, M.; Hu, S.; Meng, S.; Goulielmakis, E. Laser picoscopy of valence electrons in solids. *Nature* **2020**, *583*, 55–59. [CrossRef] [PubMed]
26. Steinmetz, S.A.; Kliewer, C.J. Gas detection sensitivity of hybrid fs/ps and fs/ns CARS. *Opt. Lett.* **2022**, *47*, 1470–1473. [CrossRef] [PubMed]
27. Hosseinnia, A.; Ruchkina, M.; Ding, P.; Bengtsson, P.-E.; Bood, J. Simultaneous temporally and spectrally resolved Raman coherences with single-shot fs/ns rotational CARS. *Opt. Lett.* **2020**, *45*, 308–311. [CrossRef]
28. Hosseinnia, A.; Ruchkina, M.; Ding, P.; Bood, J.; Bengtsson, P.-E. Single-shot fs/ns rotational CARS for temporally and spectrally resolved gas-phase diagnostics. *Proc. Combust. Inst.* **2021**, *38*, 1843–1850. [CrossRef]
29. Maker, P.D.; Terhune, R.W. Study of Optical Effects Due to an Induced Polarization Third Order in the Electric Field Strength. *Phys. Rev.* **1965**, *148*, 990. [CrossRef]
30. Regnier, P.R.; Moya, F.; Taran, J.P.E. Gas Concentration Measurement by Coherent Raman Anti-Stokes Scattering. *AIAA J.* **1974**, *12*, 826–831. [CrossRef]
31. Moya, F.; Druet, S.A.J.; Taran, J.P.E. Gas spectroscopy and temperature measurement by coherent Raman anti-stokes scattering. *Opt. Commun.* **1975**, *13*, 169–174. [CrossRef]
32. Roy, S.; Gord, J.R.; Patnaik, A.K. Recent advances in coherent anti-Stokes Raman scattering spectroscopy: Fundamental developments and applications in reacting flows. *Prog. Energ. Combust.* **2010**, *36*, 280–306. [CrossRef]
33. Gord, J.R.; Meyer, T.R.; Roy, S. Applications of ultrafast lasers for optical measurements in combusting flows. *Annu. Rev. Anal. Chem.* **2008**, *1*, 663–687. [CrossRef] [PubMed]
34. Slabaugh, C.D.; Dennis, C.N.; Boxx, I.; Meier, W.; Lucht, R.P. 5 kHz thermometry in a swirl-stabilized gas turbine model combustor using chirped probe pulse femtosecond CARS. Part 2. Analysis of swirl flame dynamics. *Combust. Flame* **2016**, *173*, 454–467. [CrossRef]
35. Lang, T.; Kompa, K.L.; Motzkus, M. Femtosecond CARS on H_2. *Chem. Phys. Lett.* **1999**, *310*, 65–72. [CrossRef]
36. Eckbreth, A.C. BOXCARS: Crossed-beam phase-matched CARS generation in gases. *Appl. Phys. Lett.* **1978**, *32*, 421–423. [CrossRef]
37. Zheng, J.B.; Leipertz, A.; Snow, J.B.; Chang, R.K. Simultaneous observation of rotational coherent Stokes Raman scattering and coherent anti-Stokes Raman scattering in air and nitrogen. *Opt. Lett.* **1983**, *8*, 350–352. [CrossRef]
38. Prior, Y. Three-dimensional phase matching in four-wave mixing. *Appl. Opt.* **1980**, *19*, 1741_1–1743. [CrossRef] [PubMed]
39. Shirley, J.A.; Hall, R.J.; Eckbreth, A.C. Folded BOXCARS for rotational Raman studies. *Opt. Lett.* **1980**, *5*, 380. [CrossRef] [PubMed]
40. Bohlin, A.; Kliewer, C.J. Direct Coherent Raman Temperature Imaging and Wideband Chemical Detection in a Hydrocarbon Flat Flame. *J. Phys. Chem. Lett.* **2015**, *6*, 643–649. [CrossRef]
41. Kearney, S.P. Hybrid fs/ps rotational CARS temperature and oxygen measurements in the product gases of canonical flat flames. *Combust. Flame* **2015**, *162*, 1748–1758. [CrossRef]
42. Engel, S.R.; Miller, J.D.; Dedic, C.E.; Seeger, T.; Leipertz, A.; Meyer, T.R. Hybrid femtosecond/picosecond coherent anti-Stokes Raman scattering for high-speed CH_4/N_2 measurements in binary gas mixtures. *J. Raman Spectrosc.* **2013**, *44*, 1336–1343. [CrossRef]
43. Miller, J.D.; Dedic, C.E.; Meyer, T.R. Vibrational femtosecond/picosecond coherent anti-Stokes Raman scattering with enhanced temperature sensitivity for flame thermometry from 300 to 2400 K. *J Raman Spectrosc* **2015**, *46*, 702–707. [CrossRef]
44. Scherman, M.; Nafa, M.; Schmid, T.; Godard, A.; Bresson, A.; Attal-Tretout, B.; Joubert, P. Rovibrational hybrid fs/ps CARS using a volume Bragg grating for N(2) thermometry. *Opt. Lett.* **2016**, *41*, 488–491. [CrossRef] [PubMed]

45. Kliewer, C.J. High-spatial-resolution one-dimensional rotational coherent anti-Stokes Raman spectroscopy imaging using counterpropagating beams. *Opt. Lett.* **2012**, *37*, 229–231. [CrossRef] [PubMed]
46. Courtney, T.L.; Mecker, N.T.; Patterson, B.D.; Linne, M.; Kliewer, C.J. Hybrid femtosecond/picosecond pure rotational anti-Stokes Raman spectroscopy of nitrogen at high pressures (1–70 atm) and temperatures (300–1000 K). *Appl. Phys. Lett.* **2019**, *114*, 101107. [CrossRef]
47. Yang, C.B.; He, P.; Escofet-Martin, D.; Peng, J.B.; Fan, R.W.; Yu, X.; Dunn-Rankin, D. Impact of input field characteristics on vibrational femtosecond coherent anti-Stokes Raman scattering thermometry. *Appl. Opt.* **2018**, *57*, 197–207. [CrossRef]
48. Welford, W.T. The Principles of Nonlinear Optics. *Phys. Bull.* **1985**, *36*, 178. [CrossRef]
49. Cheng, J.X.; Book, L.D.; Xie, X.S. Polarization coherent anti-Stokes Raman scattering microscopy. *Opt. Lett.* **2001**, *26*, 1341–1343. [CrossRef] [PubMed]
50. Cheng, J.-X.; Volkmer, A.; Xie, X.S. Theoretical and experimental characterization of coherent anti-Stokes Raman scattering microscopy. *J. Opt. Soc. Am. B* **2002**, *19*, 1363–1375. [CrossRef]
51. Dudovich, N.; Oron, D.; Silberberg, Y. Single-pulse coherently controlled nonlinear Raman spectroscopy and microscopy. *Nature* **2002**, *418*, 512–514. [CrossRef]
52. Song, Y.; Wu, H.; Zhu, G.; Zeng, Y.; Yu, G.; Yang, Y. A femtosecond time-resolved coherent anti-Stokes Raman spectroscopy thermometry for steady-state high-temperature flame. *Combust. Flame* **2022**, *242*, 112166. [CrossRef]
53. Kulatilaka, W.D.; Stauffer, H.U.; Gord, J.R.; Roy, S. One-dimensional single-shot thermometry in flames using femtosecond-CARS line imaging. *Opt. Lett.* **2011**, *36*, 4182–4184. [CrossRef] [PubMed]
54. Mazza, F.; Griffioen, N.; Castellanos, L.; Kliukin, D.; Bohlin, A. High-temperature rotational-vibrational O_2CO_2 coherent Raman spectroscopy with ultrabroadband femtosecond laser excitation generated in-situ. *Combust. Flame* **2022**, *237*, 111738. [CrossRef]
55. Zinth, W.; Laubereau, A.; Kaiser, W. Time resolved observation of resonant and non-resonant contributions to the nonlinear susceptibility $\chi(3)$. *Opt. Commun.* **1978**, *26*, 457–462. [CrossRef]
56. Hayden, C.C.; Chandler, D.W. Femtosecond time-resolved studies of coherent vibrational Raman scattering in large gas-phase molecules. *J. Chem. Phys.* **1995**, *103*, 10465–10472. [CrossRef]
57. Lang, T.; Motzkus, M.; Frey, H.M.; Beaud, P. High resolution femtosecond coherent anti-Stokes Raman scattering: Determination of rotational constants, molecular anharmonicity, collisional line shifts, and temperature. *J. Chem. Phys.* **2001**, *115*, 5418–5426. [CrossRef]
58. Beaud, P.; Frey, H.M.; Lang, T.; Motzkus, M. Flame thermometry by femtosecond CARS. *Chem. Phys. Lett.* **2001**, *344*, 407–412. [CrossRef]
59. Roy, S.; Kinnius, P.J.; Lucht, R.P.; Gord, J.R. Temperature measurements in reacting flows by time-resolved femtosecond coherent anti-Stokes Raman scattering (fs-CARS) spectroscopy. *Opt. Commun.* **2008**, *281*, 319–325. [CrossRef]
60. Roy, S.; Richardson, D.; Kinnius, P.J.; Lucht, R.P.; Gord, J.R. Effects of N_2–CO polarization beating on femtosecond coherent anti-Stokes Raman scattering spectroscopy of N_2. *Appl. Phys. Lett.* **2009**, *94*, 144101. [CrossRef]
61. Lucht, R.P.; Kinnius, P.J.; Roy, S.; Gord, J.R. Theory of femtosecond coherent anti-Stokes Raman scattering spectroscopy of gas-phase transitions. *J. Chem. Phys.* **2007**, *127*, 044316. [CrossRef]
62. Lucht, R.P.; Roy, S.; Meyer, T.R.; Gord, J.R. Femtosecond coherent anti-Stokes Raman scattering measurement of gas temperatures from frequency-spread dephasing of the Raman coherence. *Appl. Phys. Lett.* **2006**, *89*. [CrossRef]
63. Xia, Y.; Zhao, Y.; Zhang, T.; He, P.; Fan, R.; Dong, Z.; Chen, D.; Zhang, Z. Measurements of flame temperature by femtosecond CARS. *Chin. Opt. Lett.* **2012**, *10*, S13002–S313003. [CrossRef]
64. Roy, S.; Kulatilaka, W.D.; Richardson, D.R.; Lucht, R.P.; Gord, J.R. Gas-phase single-shot thermometry at 1 kHz using fs-CARS spectroscopy. *Opt. Lett.* **2009**, *34*, 3857–3859. [CrossRef] [PubMed]
65. Lang, T.; Motzkus, M. Determination of line shift coefficients with femtosecond time resolved CARS. *J. Raman Spectrosc.* **2000**, *31*, 65–70. [CrossRef]
66. Patnaik, A.K.; Roy, S.; Gord, J.R.; Lucht, R.P.; Settersten, T.B. Effects of collisions on electronic-resonance-enhanced co-herent anti-Stokes Raman scattering of nitric oxide. *J. Chem. Phys.* **2009**, *130*, 214304. [CrossRef] [PubMed]
67. Potma, E.O.; Jones, D.J.; Cheng, J.X.; Xie, X.S.; Ye, J. High-sensitivity coherent anti-Stokes Raman scattering microscopy with two tightly synchronized picosecond lasers. *Opt. Lett.* **2002**, *27*, 1168–1170. [CrossRef]
68. Lausten, R.; Smirnova, O.; Sussman, B.J.; Grafe, S.; Mouritzen, A.S.; Stolow, A. Time- and frequency-resolved coherent anti-Stokes Raman scattering spectroscopy with sub-25 fs laser pulses. *J. Chem. Phys.* **2008**, *128*, 244310. [CrossRef]
69. Volkmer, A.; Book, L.D.; Xie, X.S. Time-resolved coherent anti-Stokes Raman scattering microscopy: Imaging based on Raman free induction decay. *Appl. Phys. Lett.* **2002**, *80*, 1505–1507. [CrossRef]
70. Rodriguez, L.G.; Lockett, S.J.; Holtom, G.R. Coherent anti-stokes Raman scattering microscopy: A biological review. *Cytom. A* **2006**, *69*, 779–791. [CrossRef]
71. Evans, C.L.; Xu, X.; Kesari, S.; Xie, X.S.; Wong, S.T.; Young, G.S. Chemically-selective imaging of brain structures with CARS microscopy. *Opt. Express* **2007**, *15*, 12076–12087. [CrossRef]
72. Song, Y.; Wu, H.; Zhu, G.; Yang, Y.; Lei, Q.; Yu, G. Real-time temperature monitoring technology for dynamic combustion processes using dual-probe femtosecond CARS. *Opt. Laser. Eng.* **2024**, *175*, 108001. [CrossRef]
73. Dedic, C.E.; Miller, J.D.; Meyer, T.R. Dual-pump vibrational/rotational femtosecond/picosecond coherent anti-Stokes Raman scattering temperature and species measurements. *Opt. Lett.* **2014**, *39*, 6608–6611. [CrossRef] [PubMed]

74. Richardson, D.R.; Lucht, R.P.; Kulatilaka, W.D.; Roy, S.; Gord, J.R. Theoretical modeling of single-laser-shot, chirped-probe-pulse femtosecond coherent anti-Stokes Raman scattering thermometry. *Appl. Phys. B* **2011**, *104*, 699–714. [CrossRef]
75. Lang, T.; Motzkus, M. Single-shot femtosecond coherent anti-Stokes Raman-scattering thermometry. *J. Opt. Soc. Am. B* **2002**, *19*, 340–344. [CrossRef]
76. Richardson, D.R.; Lucht, R.P.; Roy, S.; Kulatilaka, W.D.; Gord, J.R. Single-laser-shot femtosecond coherent anti-Stokes Raman scattering thermometry at 1000Hz in unsteady flames. *Proc. Combust. Inst.* **2011**, *33*, 839–845. [CrossRef]
77. Richardson, D.R.; Bangar, D.; Lucht, R.P. Polarization suppression of the nonresonant background in femtosecond coherent anti-Stokes Raman scattering for flame thermometry at 5 kHz. *Opt. Express* **2012**, *20*, 21495–21504. [CrossRef] [PubMed]
78. Tolles, W.M.; Nibler, J.W.; McDonald, J.R.; Harvey, A.B. A Review of the Theory and Application of Coherent Anti-Stokes Raman Spectroscopy (CARS). *Appl. Spectrosc.* **1977**, *31*, 253–271. [CrossRef]
79. Dennis, C.N.; Cruise, D.L.; Mongia, H.C.; King, G.B.; Lucht, R.P. Study of Swirl Stabilized Burner with Interchangeable Swirler Using Chirped-Probe-Pulse Femtosecond Coherent Anti-Stokes Raman Scattering for Thermometry and CH4 Concentration Measurements. In Proceedings of the AIAA SciTech Forum, Kissimmee, FL, USA & Online, 5–9 January 2015. [CrossRef]
80. Fineman, C.N.; Lucht, R.P. Sooting Jet Diffusion Flame Thermometry at 5 kHz using Femtosecond Coherent Anti-Stokes Raman Scattering. *AIAA J.* **2014**, *3*, 3731–3738. [CrossRef]
81. Dennis, C.N.; Slabaugh, C.D.; Boxx, I.G.; Meier, W.; Lucht, R.P. Chirped probe pulse femtosecond coherent anti-Stokes Raman scattering thermometry at 5 kHz in a Gas Turbine Model Combustor. *Proc. Combust. Inst.* **2015**, *35*, 3731–3738. [CrossRef]
82. Dennis, C.N.; Slabaugh, C.D.; Boxx, I.G.; Meier, W.; Lucht, R.P. 5 kHz thermometry in a swirl-stabilized gas turbine model combustor using chirped probe pulse femtosecond CARS. Part 1: Temporally resolved swirl-flame thermometry. *Combust. Flame* **2016**, *173*, 441–453. [CrossRef]
83. Thomas, L.M.; Satija, A.; Lucht, R.P. Technique developments and performance analysis of chirped-probe-pulse femtosecond coherent anti-Stokes Raman scattering combustion thermometry. *Appl. Opt.* **2017**, *56*, 8797–8810. [CrossRef] [PubMed]
84. Lowe, A.; Thomas, L.M.; Satija, A.; Lucht, R.P.; Masri, A.R. Chirped-probe-pulse femtosecond CARS thermometry in turbulent spray flames. *Proc. Combust. Inst.* **2019**, *37*, 1383–1391. [CrossRef]
85. Lowe, A.; Thomas, L.M.; Satija, A.; Lucht, R.P.; Masri, A.R. Five kHz thermometry in turbulent spray flames using chirped-probe-pulse femtosecond CARS, part II: Structure of reaction zones. *Combust. Flame* **2019**, *200*, 417–432. [CrossRef]
86. Su, J.; Xie, R.; Johnson, C.K.; Hui, R. Single fiber laser based wavelength tunable excitation for CRS spectroscopy. *J. Opt. Soc. Am. B* **2013**, *30*, 1671–1682. [CrossRef] [PubMed]
87. Rocha-Mendoza, I.; Langbein, W.; Borri, P. Coherent anti-Stokes Raman microspectroscopy using spectral focusing with glass dispersion. *Appl. Phys. Lett.* **2008**, *93*, 201103. [CrossRef]
88. Richardson, D.R.; Stauffer, H.U.; Roy, S.; Gord, J.R. Comparison of chirped-probe-pulse and hybrid femtosecond/picosecond coherent anti-Stokes Raman scattering for combustion thermometry. *Appl. Opt.* **2017**, *56*, E37–E49. [CrossRef]
89. Zhao, H.; Tian, Z.; Wu, T.; Li, Y.; Wei, H. Optimization of probe time delays in hybrid femtosecond/picosecond vibrational coherent anti-Stokes Raman scattering thermometry. In Proceedings of the CLEO: Science and Innovations, San Jose, CA, USA, 10–15 May 2020; Optica Publishing Group: Washington, DC, USA, May 2020; p. JW2F-15. [CrossRef]
90. Prince, B.D.; Chakraborty, A.; Prince, B.M.; Stauffer, H.U. Development of simultaneous frequency- and time-resolved coherent anti-Stokes Raman scattering for ultrafast detection of molecular Raman spectra. *J. Chem. Phys.* **2006**, *125*, 044502. [CrossRef]
91. Miller, J.D.; Slipchenko, M.N.; Meyer, T.R.; Stauffer, H.U.; Gord, J.R. Hybrid femtosecond/picosecond coherent anti-Stokes Raman scattering for high-speed gas-phase thermometry. *Opt. Lett.* **2010**, *35*, 2430–2432. [CrossRef] [PubMed]
92. Miller, J.D.; Roy, S.; Slipchenko, M.N.; Gord, J.R.; Meyer, T.R. Single-shot gas-phase thermometry using pure-rotational hybrid femtosecond/picosecond coherent anti-Stokes Raman scattering. *Opt. Express* **2011**, *19*, 15627–15640. [CrossRef]
93. Escofet-Martin, D.; Ojo, A.O.; Collins, J.; Mecker, N.T.; Linne, M.; Peterson, B. Dual-probe 1D hybrid fs/ps rotational CARS for simultaneous single-shot temperature, pressure, and O_2/N_2 measurements. *Opt. Lett.* **2020**, *45*, 4758–4761. [CrossRef]
94. Kim, A.; Dedic, C.E.; Cutler, A.D. Development of a fs/ps CARS system for temperature and species measurements in a dual-mode scramjet combustor. In Proceedings of the AIAA SCITECH 2023 Forum, National Harbor, MD, USA & Oline, 23–27 January 2023. [CrossRef]
95. Seeger, T.; Leipertz, A. Experimental comparison of single-shot broadband vibrational and dual-broadband pure rotational coherent anti-Stokes Raman scattering in hot air. *Appl. Opt.* **1996**, *35*, 2665–2671. [CrossRef] [PubMed]
96. Thumann, A.; Schenk, M.; Jonuscheit, J.; Seeger, T.; Leipertz, A. Simultaneous temperature and relative nitrogen–oxygen concentration measurements in air with pure rotational coherent anti-Stokes Raman scattering for temperatures to as high as 2050 K. *Appl. Opt.* **1997**, *36*, 3500–3505. [CrossRef] [PubMed]
97. Kearney, S.P.; Scoglietti, D.J.; Kliewer, C.J. Hybrid femtosecond/picosecond rotational coherent anti-Stokes Raman scattering temperature and concentration measurements using two different picosecond-duration probes. *Opt. Express* **2013**, *21*, 1232–12339. [CrossRef] [PubMed]
98. Bohlin, A.; Kliewer, C.J. Communication: Two-dimensional gas-phase coherent anti-Stokes Raman spectroscopy (2D-CARS): Simultaneous planar imaging and multiplex spectroscopy in a single laser shot. *J. Chem. Phys.* **2013**, *138*, 221101. [CrossRef] [PubMed]

99. Stauffer, H.U.; Miller, J.D.; Roy, S.; Gord, J.R.; Meyer, T.R. Communication: Hybrid femtosecond/picosecond rotational coherent anti-Stokes Raman scattering thermometry using a narrowband time-asymmetric probe pulse. *J. Chem. Phys.* **2012**, *136*, 111101. [CrossRef] [PubMed]
100. Yang, C.; Escofet-Martin, D.; Dunn-Rankin, D.; Chien, Y.-C.; Yu, X.; Mukamel, S. Hybrid femtosecond/picosecond pure-rotational coherent anti-Stokes Raman scattering with chirped probe pulses. *J. Raman Spectrosc.* **2017**, *48*, 1881–1886. [CrossRef]
101. Zhao, H.; Tian, Z.; Wu, T.; Li, Y.; Wei, H. Dynamic and sensitive hybrid fs/ps vibrational CARS thermometry using a quasi-common-path second-harmonic bandwidth-compressed probe. *Appl. Phys. Lett.* **2021**, *118*, 071107. [CrossRef]
102. Cutler, A.D.; Cantu, L.M.L.; Gallo, E.C.A.; Baurle, R.; Danehy, P.M.; Rockwell, R.; Goyne, C.; McDaniel, J. Nonequilibrium Supersonic Freestream Studied Using Coherent Anti-Stokes Raman Spectroscopy. *AIAA J.* **2015**, *53*, 2762–2770. [CrossRef]
103. Magnotti, G.; Cutler, A.D.; Herring, G.C.; Tedder, S.A.; Danehy, P.M. Saturation and Stark broadening effects in dual-pump CARS of N_2, O_2, and H_2. *J. Raman Spectrosc.* **2012**, *43*, 611–620. [CrossRef]
104. Wrzesinski, P.J.; Stauffer, H.U.; Kulatilaka, W.D.; Gord, J.R.; Roy, S. Time-resolved femtosecond CARS from 10 to 50 Bar: Collisional sensitivity. *J. Raman Spectrosc.* **2013**, *44*, 1344–1348. [CrossRef]
105. Stauffer, H.U.; Rahman, K.A.; Slipchenko, M.N.; Roy, S.; Gord, J.R.; Meyer, T.R. Interference-free hybrid fs/ps vibrational CARS thermometry in high-pressure flames. *Opt. Lett.* **2018**, *43*, 4911–4914. [CrossRef]
106. Vestin, F.; Afzelius, M.; Bengtsson, P.E. Improved species concentration measurements using a species-specific weighting procedure on rotational CARS spectra. *J. Raman Spectrosc.* **2005**, *36*, 95–101. [CrossRef]
107. Rockwell, R.D.; Goyne, C.P.; Haw, W.; Krauss, R.H.; McDaniel, J.C.; Trefny, C.J. Experimental Study of Test-Medium Vitiation Effects on Dual-Mode Scramjet Performance. *J. Propul. Power* **2011**, *27*, 1135–1142. [CrossRef]
108. Rockwell, R.D.; Goyne, C.P.; Rice, B.E.; Kouchi, T.; McDaniel, J.C.; Edwards, J.R. Collaborative Experimental and Computational Study of a Dual-Mode Scramjet Combustor. *J. Propul. Power* **2014**, *30*, 530–538. [CrossRef]
109. Cutler, A.D.; Magnotti, G.; Cantu, L.; Gallo, E.; Rockwell, R.; Goyne, C. Dual-Pump Coherent Anti-Stokes Raman Spectroscopy Measurements in a Dual-Mode Scramjet. *J. Propul. Power* **2014**, *30*, 539–549. [CrossRef]
110. Cutler, A.D.; Gallo, E.C.A.; Cantu, L.M.L.; Rockwell, R.D.; Goyne, C.P. Coherent anti-Stokes Raman spectroscopy of a premixed ethylene–air flame in a dual-mode scramjet. *Combust. Flame* **2018**, *189*, 92–105. [CrossRef]

photonics

MDPI

Article

Experimental Investigations and Modeling of Interference Fringe Geometry in Line-Shaped Gaussian Beam Intersections for Laser Doppler Sensors

Hao Zhang *, Jieping Wang and Shiji Wang

Department of Measurement and Control Technology and Instrumentation, Dalian Maritime University, Dalian 116026, China

* Correspondence: hao.zhang@dlmu.edu.cn

Abstract: Line-shaped beam-based Doppler distance sensors enable 3D shape measurements of rotating objects, for instance for working lathe monitoring with a simultaneous, multipoint velocity, and distance measurement. The velocity and distance estimations are fundamentally referred to the interference fringe spacing of the sensors. In this case, the fringe spacing variation-caused measurement error is significant; however, a complete, accurate model of the fringe geometry distribution for the line-shaped fringe volume is still missing and required to identify and minimize the error. Therefore, this work presents a high-accuracy 3D model for the fringe spacing evaluation of the sensors with experimental investigations. The model is derived from the phase expression of Gaussian beams introducing extension ratio, and can be universally applied to describe fringe geometry distribution throughout the intersection volume of spherical and line-shaped beams. With an experimental setup of a laser Doppler sensor, a full-field fringe spacing estimation using a high-resolution matrix camera is performed with dual-wavelength beams. The numerical modeling and experimental results show an average relative difference below 0.6%.

Keywords: interferometry; fringe geometry; 3D modeling; beam shaping; laser Doppler

1. Introduction

The simultaneous optical in-process measurement of the velocity and distance of moving objects is a significant task in diverse applications, such as moving rough surfaces and tracer particles' monitoring for workpiece shape measurements in turning lathes and flow phenomena detections of microfluidics. For this purpose, a novel laser Doppler distance sensor is developed [1–3]. Compared to conventional optical distance measurement techniques [4–11], it offers the advantage of simultaneous axial distance and lateral velocity measurement based on two interference fringe systems by superimposing dual-wavelength Gaussian beams. In this sensor, the velocity and distance are respectively determined from the Doppler frequency and phase difference of the speckle signals generated by optically rough surfaces or particles passing through the fringe systems. The Doppler frequency and phase difference of the speckle signals are fundamentally referred to the fringe spacing [12]. Therefore, in order to obtain an accurate measurement, a significant step is to achieve accurate, full-field fringe spacing distribution in the intersection volume.

With the birth of lasers [13,14], the laser beam with a Gaussian intensity profile is perhaps the most important one, which is often called the Gaussian beam and the fundamental mode as compared to the higher order modes [15]. The Gaussian beam is widely used in the research and application of laser devices [16,17], optical processing, and measurement [18,19]. The Gaussian intensity profile of the beams is determined by diffraction effects. Historically, the modes were approximated by wave beams, and the concept of electromagnetic wave beams was introduced by investigating the properties of sequences of lenses for the transmission of electromagnetic waves. The resonant properties of Gaussian beams in the resonator structure, the propagation characteristics in free space, and the behaviors as they interact with diverse optical systems have been investigated since the 1960s [15]. As limitations of the early research, the investigations of Gaussian beams mainly focused on the passage of paraxial rays through optical elements, the wave nature of the beams, and diffraction effects. With the development of laser measurement techniques, the study of Gaussian beams needs to be combined with practical applications. Especially for the dual-beam laser Doppler sensing, the influence of the properties of Gaussian beams on the interference field urgently needs to be investigated and quantified for improving measurement accuracy.

In dual Gaussian beam interference, the effect of beam nature on the variation of fringe spacing in the intersection volume is significant due to its potential impact on the applications. In the initial fringe field analyses of the intersection volume, the valid expressions were derived for the variation in fringe spacing along the probe volume longitudinal axis and along the transverse axis perpendicular to the fringes [20]. An alternative expression is deduced for the transverse variation and the longitudinal model is extended by referring the results to system parameters [21]. The resulting equations are effective for evaluating the longitudinal fringe spacing only when the beam waists are far from the center of intersection volume. They are not valid under a relatively good alignment, which is the condition of greatest interest and most uses. The above works all contain multiple approximations whose effects are hard to quantify. In another way, an indirect fringe distortion inspection by using signal frequency error analysis was proposed considering various system parameters such as the beam crossing angle and the lens focal length [22]. But the formulas obtained are not feasible with simple computation and the synergism of these parameters on the fringe variation are still unclear. With further improvement, the valid expressions of fringe spacing throughout the intersection volume of two Gaussian beams were performed, which can be simplified by means of precisely quantified approximations for an easy calculation [23]. But the expressions describe the longitudinal or transverse variation of fringe geometry independently and are only applicable to spherical Gaussian beams. Up to now, a universal, accurate 3D model of the fringe spacing distribution in the intersection volume is still missing and required. Especially for the laser Doppler distance sensor with line-shaped beam based multipoint measurement [24], the fringe geometry change along the height of the intersection volume leads to a great systematic error and therefore needs to be exactly investigated and computed to eliminate the error.

The aim of this paper is to present an exact, universal model of fringe spacing distribution for calibrating the interference fringe field in paraxial Gaussian beam intersections. A comprehensive analysis resulting in accurate 3D expressions for the fringe spacing is conducted first in Section 2. The expressions refer to system parameters relevant to the optics configuration, enabling a priori evaluation of fringe variation for system optimization. To demonstrate the model, an experimental setup of the laser Doppler system is developed and a full-field fringe spacing evaluation is performed with line-shaped beams in two wavelengths. With the consistent condition, the numerical modeling of fringe spacing is conducted and the relative differences of the modeling and experimental results are investigated in Section 3.

2. Model of Fringe Geometry

The dual-beam mode is an optical method that is widely used in laser Doppler sensors for generating Doppler frequency-modulated scattered light signals, cf. Figure 1. In this mode, two Gaussian beams from coherent light sources are superimposed to form an interference fringe volume, which offers a fundamental dimension for the following analysis of fringe spacing. For a global, intuitive observation and analysis, the coordinate transformation from the independent coordinates of the two beams to the beam intersection volume coordinate is essential.

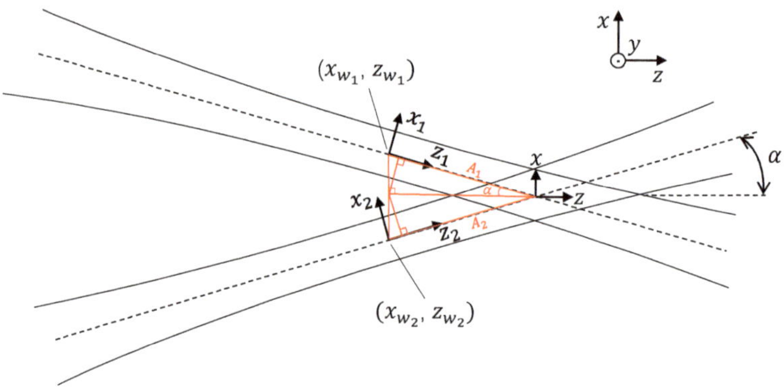

Figure 1. Scheme of geometry and coordinate system with dual-beam mode for the analysis of fringe spacing.

In the intersection volume, the interference is generated by two beams crossing with a half angle α. x_i, y_i, and z_i are defined as coordinates in the beam coordinate system with its origin at the waist of beam i. The position in the range of the intersection volume is described by the x, y, z coordinate system, the origin of which is at the intersection of the two beam centerlines. x_{w_i} and z_{w_i} are the waist coordinates of beam i in the x, y, z coordinate system, which are also the origin coordinates of the x_i, y_i, z_i coordinate system and constants for a certain dual-beam optical system. Based on the geometric relation, $A_1 = -z_{w_1} \cdot \cos\alpha + x_{w_1} \cdot \sin\alpha$, $A_2 = -z_{w_2} \cdot \cos\alpha - x_{w_2} \cdot \sin\alpha$. The transformation between x_i, y_i, z_i and x, y, z for two beams can thus be achieved by rotation and translation:

$$\begin{bmatrix} x_1 \\ y_1 \\ z_1 \\ 1 \end{bmatrix} = \begin{bmatrix} 1 & 0 & 0 & 0 \\ 0 & 1 & 0 & 0 \\ 0 & 0 & 1 & A_1 \\ 0 & 0 & 0 & 1 \end{bmatrix} \cdot \begin{bmatrix} \cos\alpha & 0 & \sin\alpha & 0 \\ 0 & 1 & 0 & 0 \\ -\sin\alpha & 0 & \cos\alpha & 0 \\ 0 & 0 & 0 & 1 \end{bmatrix} \cdot \begin{bmatrix} x \\ y \\ z \\ 1 \end{bmatrix},$$
$$\begin{bmatrix} x_2 \\ y_2 \\ z_2 \\ 1 \end{bmatrix} = \begin{bmatrix} 1 & 0 & 0 & 0 \\ 0 & 1 & 0 & 0 \\ 0 & 0 & 1 & A_2 \\ 0 & 0 & 0 & 1 \end{bmatrix} \cdot \begin{bmatrix} \cos\alpha & 0 & -\sin\alpha & 0 \\ 0 & 1 & 0 & 0 \\ \sin\alpha & 0 & \cos\alpha & 0 \\ 0 & 0 & 0 & 1 \end{bmatrix} \cdot \begin{bmatrix} x \\ y \\ z \\ 1 \end{bmatrix}, \quad (1)$$

which then offer the coordinate relation:

$$\begin{aligned} x_1 &= x\cos\alpha + z\sin\alpha,\ y_1 = y, \\ z_1 &= -(x - x_{w_1})\sin\alpha + (z - z_{w_1})\cos\alpha, \\ x_2 &= x\cos\alpha - z\sin\alpha,\ y_2 = y, \\ z_2 &= (x - x_{w_2})\sin\alpha + (z - z_{w_2})\cos\alpha. \end{aligned} \quad (2)$$

By the expression of the radiation field of Gaussian beams [25], the phase variation per meter $k = 2\pi/\lambda$ and the beam Rayleigh range $z_{R_i} = \pi w_{0_i}^2/\lambda$, the phase of each beam is given as

$$\theta_i(x, y, z) = -kz_i + \tan^{-1}\frac{z_i}{z_{R_i}} - \frac{k}{2}\frac{x_i^2 + y_i^2}{R(z_i)}. \quad (3)$$

The minimum radius of the beam at the beam waist is given by w_{0_i}. In Equation (3), the first two factors describe the phase propagation towards z_i. The last factor indicates the dependency of the phase on the lateral coordinates x_i and y_i as well as the phase curvature radius $R(z_i)$, which depends on z_i:

$$R(z_i) = z_i\left[1 + \left(\frac{z_{R_i}}{z_i}\right)^2\right], \tag{4}$$

and the beam spot size is

$$w(z_i) = w_{0_i}\sqrt{1 + \left(\frac{z_i}{z_{R_i}}\right)^2}. \tag{5}$$

Interference fringes in the scope of intersection volume originate from a constant phase difference and, thus, are expressed as

$$\begin{aligned}2n\pi = \theta_1(x,y,z) - \theta_2(x,y,z) =& 2\pi\left[-\frac{1}{\lambda}(z_1 - z_2) + \frac{1}{2\pi}\left(\tan^{-1}\frac{z_1}{z_{R_1}} - \tan^{-1}\frac{z_2}{z_{R_2}}\right)\right. \\ &\left.- \frac{1}{2\lambda}\left(\frac{x_1^2 + y_1^2}{R(z_1)} - \frac{x_2^2 + y_2^2}{R(z_2)}\right)\right].\end{aligned} \tag{6}$$

Substituting Equations (2) and (4) into Equation (6) and differentiating with respect to x, the fringe spacing L yields

$$\begin{aligned}L = \left(\frac{dn}{dx}\right)^{-1} =& \frac{\lambda}{2\sin\alpha}\left[1 - \frac{1}{2\tan\alpha}\left(\frac{x_1 z_1}{z_1{}^2 + z_{R_1}{}^2} - \frac{x_2 z_2}{z_2{}^2 + z_{R_2}{}^2}\right) - \frac{z_{R_1}\lambda}{4\pi(z_1{}^2 + z_{R_1}{}^2)}\right. \\ &- \frac{z_{R_2}\lambda}{4\pi(z_2{}^2 + z_{R_2}{}^2)} - \frac{x_1^2(z_1{}^2 - z_{R_1}{}^2)}{4(z_1{}^2 + z_{R_1}{}^2)^2} - \frac{x_2^2(z_2{}^2 - z_{R_2}{}^2)}{4(z_2{}^2 + z_{R_2}{}^2)^2} \\ &\left.- \frac{y_1^2(z_1{}^2 - z_{R_1}{}^2)}{4(z_1{}^2 + z_{R_1}{}^2)^2} - \frac{y_2^2(z_2{}^2 - z_{R_2}{}^2)}{4(z_2{}^2 + z_{R_2}{}^2)^2}\right]^{-1}.\end{aligned} \tag{7}$$

Since Equation (6) is a composite function, it is inevitable to reintroduce x_i and z_i during the derivation. A derivation process is offered by Equations (A1)–(A3) in Appendix A. For a relatively concise expression, parameter replacement is not performed again for Equation (7). In most computer modeling conditions, combining Equation (2), the variables can be simply and quickly transformed.

Terms $\frac{x_i^2(z_i{}^2 - z_{R_i}{}^2)}{4(z_i{}^2 + z_{R_i}{}^2)^2}$ and $\frac{y_i^2(z_i{}^2 - z_{R_i}{}^2)}{4(z_i{}^2 + z_{R_i}{}^2)^2}$ are derived from the transverse phase terms in Equation (6) with respect to the phase curvature radius $R(z_i)$. The term $\frac{z_{R_i}\lambda}{4\pi(z_i{}^2 + z_{R_i}{}^2)}$ is associated with the Guoy phase shift $\tan^{-1}\frac{z_i}{z_{R_i}}$. As the added phase shift is greatest around the beam waist, term $\frac{z_{R_i}\lambda}{4\pi(z_i{}^2 + z_{R_i}{}^2)}$ is maximized under ideal alignment conditions when $z_i \approx 0$. By substituting $z_i = 0$ into the terms, it is seen that the maximum magnitude of the terms is equal to $\lambda^2/(2\pi w_{0_i})^2$. In order for these terms to contribute more than 0.01% to the fringe spacing, the beam waists must be less than 10 µm with $\lambda = 685$ nm and less than 7.5 µm with $\lambda = 500$ nm. Under most practical circumstances, the $\frac{z_{R_i}\lambda}{4\pi(z_i{}^2 + z_{R_i}{}^2)}$ terms can thus be neglected. Regarding $\frac{x_i^2(z_i{}^2 - z_{R_i}{}^2)}{4(z_i{}^2 + z_{R_i}{}^2)^2}$ terms, the maximum of x_i^2 is determined by the square of beam radius $w^2(z_i)$, i.e., $w_{0_i}^2(1 + \frac{z_i^2}{z_{R_i}^2})$. With $z_{R_i} = \pi w_{0_i}^2/\lambda$, the maximum of $\frac{x_i^2(z_i{}^2 - z_{R_i}{}^2)}{4(z_i{}^2 + z_{R_i}{}^2)^2}$ equals to $\left(\frac{\lambda}{2\pi w_{0_i}}\right)^2 \cdot \frac{z_i{}^2 - z_{R_i}{}^2}{z_i{}^2 + z_{R_i}{}^2}$, which is bounded by $\lambda^2/(2\pi w_{0_i})^2$ as well, and therefore can be neglected.

For the laser Doppler sensor with line-shaped Gaussian beams and camera-based detection [12,24], the intersection volume is expended and sliced towards the y-axis for high-resolution, simultaneous multipoint measurement instead of single point measurement.

Since the fringe spacing is variable for different measurement points along the y-axis, the fringe spacing variation enables a measurement uncertainty of a micron in a 3D shape measurement of a rotating object by uncertainty propagation, even if the relative variation is in the level of 10^{-3}. Thus, the fringe spacing in each measurement point must be evaluated independently. The y-axis fringe spacing variation is therefore non-ignorable and needs to be exactly investigated and calculated to minimize the systematic error. In this case, as the solo terms depending on the y coordinate, the $\frac{y_i^2(z_i{}^2-z_{R_i}{}^2)}{4(z_i{}^2+z_{R_i}{}^2)^2}$ terms are significant for describing the fringe spacing. The radius of the beam waist is defined on the x- and y-axes by w_{0x_i} and w_{0y_i}, respectively. In the interference region of the dual Gaussian beam mode, the fringe spacing is determined by the phase difference of the beams. Since the incident angle is in the x–z plane, the fringe spacing significantly depends on the phase difference along the x direction (fringe direction) and the z direction (optical axis). In this case, in the y direction, the beam waist and Rayleigh range variations, as well as the changes in phase curvature radius caused by them, have less to no influence on the fringe spacing compared to the spherical beam condition. Thus, introducing the y-axis extension ratio of the intersection volume $m = w_{0y_i}/w_{0x_i}$, and considering $y_1 = y_2 = y$, the Rayleigh range in the x direction $z_{Rx_i} = z_{R_i}$, a 3D model of the fringe spacing can be expressed as

$$L = \frac{\lambda}{2\sin\alpha}\left[1 - \frac{1}{2\tan\alpha}\left(\frac{x_1 z_1}{z_1{}^2 + z_{Rx_1}{}^2} - \frac{x_2 z_2}{z_2{}^2 + z_{Rx_2}{}^2}\right) - \frac{1}{4}\left(\frac{y}{m}\right)^2\left(\frac{z_1{}^2 - z_{Rx_1}{}^2}{(z_1{}^2 + z_{Rx_1}{}^2)^2} + \frac{z_2{}^2 - z_{Rx_2}{}^2}{(z_2{}^2 + z_{Rx_2}{}^2)^2}\right)\right]^{-1}. \quad (8)$$

Employing Equation (8), the fringe geometry throughout the intersection volume of spherical beams and line-shaped beams along the y-axis can be evaluated for arbitrary sizes and positions of the two beam waists.

3. Results and Discussion

In the experimental investigations of the fringe field, a line-shaped beam-based laser Doppler velocity and distance sensor system is developed based on a Mach–Zehnder velocimeter [26,27] with beams of dual wavelength, cf. Figure 2.

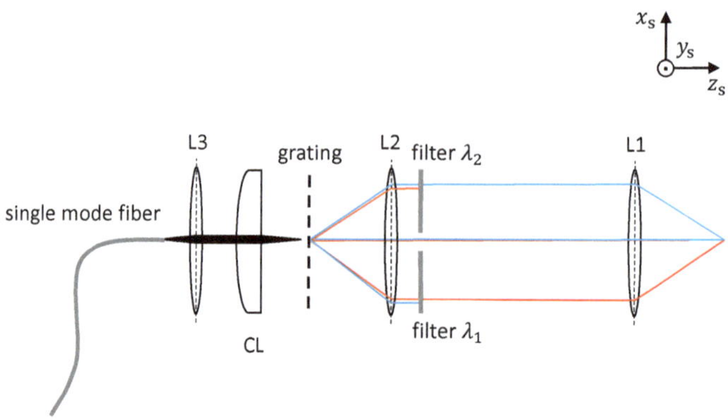

Figure 2. Cylindrical lens based sensor setup for simultaneous multipoint measurements: superposition of two interference fringe systems with fringe spacing L. The incident laser beams are not limited by the cylindrical lens CL on the y-axis. Thereby, the fringe volume is line-shaped and expanded towards the y-axis for multipoint measurements.

The laser light with the wavelengths of λ_1 = 685 nm and λ_2 = 659 nm generated by two fiber-coupled laser diodes are coupled into a single mode fiber. Through the collimating lens L3, on the distal end of the fiber, the collimated beams are focused by the cylindrical lens CL with the focal lengths $f_{cx} = 50$ mm and $f_{cy} = \infty$. Thereby, the beams remain parallel on the y-axis. A transmission diffraction grating with the grating constant $g = 4$ µm is positioned at the focal plane of CL and L2 to split the bichromatic light. One filter blocks the +1. diffraction order (DO) from λ_1 and the other one blocks the -1. DO from λ_2. The x-axis dimension of the 0. DO beams is about 25 µm, and the gap of 1 mm between the two filters is made to allow the beams crossing. Passing the Keplerian telescope consisting of lens L1 and L2 ($f_1 = 80$ mm, $f_2 = 40$ mm), the remaining beams of ± 1. and 0. DO are superimposed in the focal plane of L1 with a constant dimension on the y-axis. This produces two interference fringe systems around the beam waists with $L \approx 8$ µm, $2 \cdot w_{0y_i} \approx 2$ mm, $2 \cdot w_{0x_i} \approx 50$ µm and, thus, m is about 40. The beam waist dimensions are obtained by using camera detection and geometric optics of the system parameters. This offers a condition of good alignment with the two beam waists located at the center of the intersection volume, i.e., $x_{w_1} = x_{w_2} = 0$ and $z_{w_1} = z_{w_2} = 0$, which is of most use. By using full-field, camera-based scattered light detection, the sensor allows a simultaneous measurement of up to several hundred points. Since the half angle $\alpha = \frac{1}{2} \arctan \frac{f_2 \lambda}{f_1 g}$ and is small (below 5°), $\sin \alpha \approx \tan \alpha$ and Equation (8) can be written as

$$L = \frac{f_1 g}{f_2} \left[1 - \frac{f_1 g}{f_2 \lambda} \left(\frac{x_1 z_1}{{z_1}^2 + {z_{Rx_1}}^2} - \frac{x_2 z_2}{{z_2}^2 + {z_{Rx_2}}^2} \right) - \frac{1}{4} \left(\frac{y}{m} \right)^2 \left(\frac{{z_1}^2 - {z_{Rx_1}}^2}{({z_1}^2 + {z_{Rx_1}}^2)^2} + \frac{{z_2}^2 - {z_{Rx_2}}^2}{({z_2}^2 + {z_{Rx_2}}^2)^2} \right) \right]^{-1}. \tag{9}$$

The fringe spacing varies with the change of position in the intersection volume due to the nature of Gaussian beams. For realizing a full-field fringe geometry evaluation of the extended intersection volume, a matrix camera (UI-1492LE, resolution H × V = 3840 × 2748 pixel, pixel size $D_p = 1.67$ µm) is integrated onto a linear motion stage (MICOS LS-65, resolution 0.2 µm) in front of the sensor system to detect the fringe pattern, cf. Figure 3. The modeling and data processing in this paper are conducted by using MATLAB.

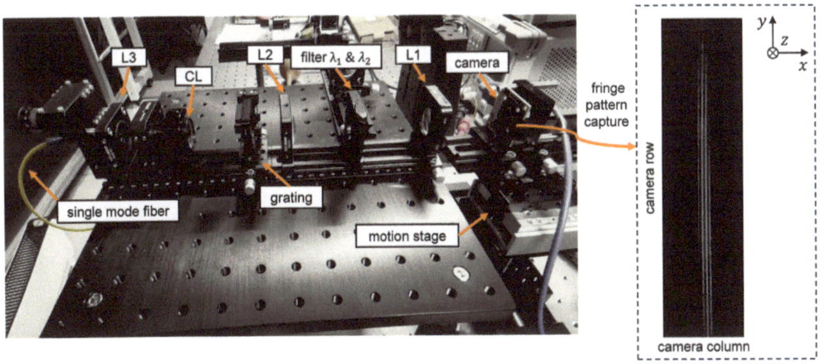

Figure 3. Experimental setup for the fringe field analysis.

The position of the camera towards the z-axis changes with the movement of the stage in a constant step size of 10 µm. Note that the x, y, z coordinate system of the beam intersection volume is different from the sensor coordinate system of x_s, y_s, z_s. The fringe pattern in each step is measured 10 times and is utilized to evaluate the fringe spacing in the amplitude spectrum employing the Fast Fourier Transform (FFT). Fringe spacing $L = \frac{1}{f_f} \cdot D_p$ is then obtained based on the fringe number per pixel f_f. The average relative measurement uncertainty of fringe spacing 0.34% is obtained. According to the measurement principle of the laser Doppler sensors, the fringe spacings along the x-axis are averaged by the FFT for estimating the fringe spacings of y- and z-axes. The fringe patterns measured in the experiment and the fringe field processing along the y-axis are shown in Figure 4.

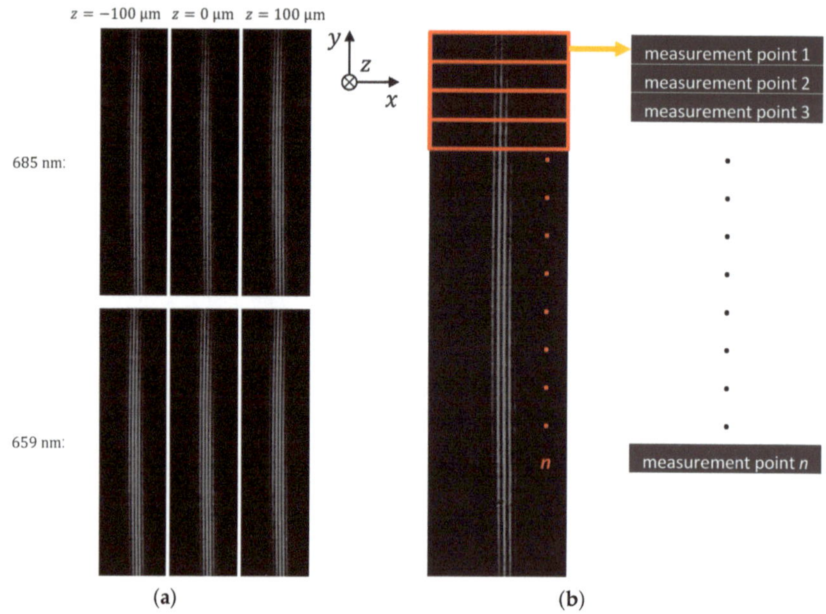

Figure 4. (**a**) The interference fringe patterns detected in the experiment. (**b**) Scheme of fringe spacing evaluation on the y-axis.

Figure 4a illustrates the fringe patterns achieved at the beam waist and the edge of the measurement range with the two wavelengths of 685 nm and 659 nm. The fringe field is sliced along the y-axis in pixels depicted in Figure 4b, which can be used for a simultaneous multipoint measurement. The fringe spacings are thus experimentally evaluated at various positions of the intersection volume in the range of 200 µm on the z-axis and 2 mm on the y-axis. Under the same conditions, the numerical modeling of fringe spacing based on Equation (9) is performed and compared with the experimental results in Figure 5.

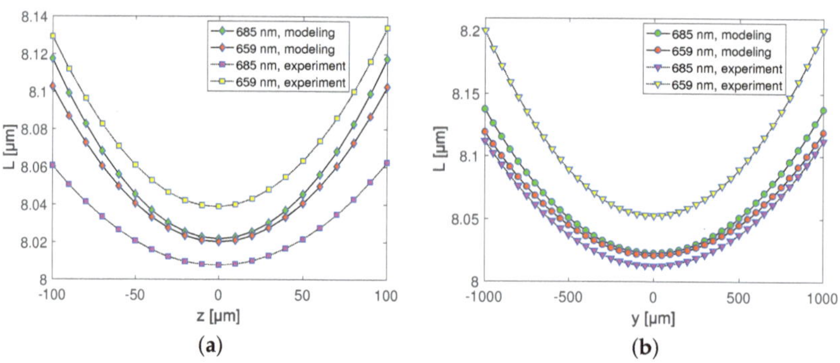

Figure 5. Fringe spacings by the modeling and experiment on the z-axis (**a**) and y-axis (**b**).

It can be seen that the trends of fringe spacings in the modeling and experiment are comparable. As shown in Figure 5a, the fringe spacings of both wavelengths almost keep consistent in the range of −30 μm to 30 μm on the *z*-axis around the center of the measurement volume, which is the position of the beam waists and indicates that the sensor system has a good alignment. It shows that the fringe spacing varies in the range of 8.02 to 8.12 μm in the modeling. At the edges, the fringe spacing differences of about 0.02 μm are indicated between the two wavelengths. For the experimental results, the fringe spacing changes from 8.01 to 8.13 μm, which is similar to the modeling results, and the maximum difference of around 0.07 μm between both wavelengths is revealed. In Figure 5b, the fringe spacing along the *y*-axis illustrates a variation from 8.02 to 8.14 μm with the modeling. The fringe spacing also shows the differences of about 0.02 μm at the edges in the both wavelengths. In the experiment, the fringe spacing is in a variation range of 8.01 to 8.2 μm and the maximum difference is about 0.08 μm between the two wavelengths.

In the modeling, a fringe spacing difference of about 0.02 μm from $\lambda/2\sin\alpha$ around the center of beam waist results from the *x*-axis averaging. The full-field maps of fringe spacings are obtained in Figure 6.

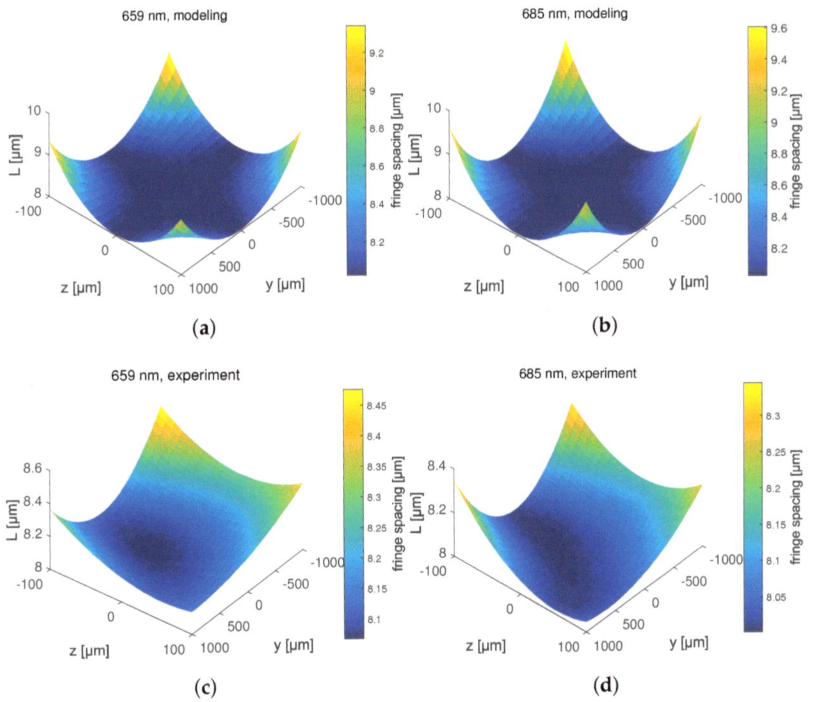

Figure 6. Full–field distribution map of fringe spacings. The modeling fringe spacing distributions with the wavelengths of 659 nm and 685 nm are depicted in (**a–d**) are the experimental fringe spacing distributions of the both wavelengths. Note the differences of the numerical scale of *L* in each graph when comparing.

All the modeling and experimental results manifest a homologous, inverted Gaussian-like distribution in which the fringe spacings around the bottom of the map are about 8 μm. In the modeling, the fringe spacing can gradually increase to 9.3 μm approaching the corner of the map. With the same coordinate range, the experimental results can reach 8.5 μm. This is comparable with the Gaussian intensity distribution and, thus, suggests that the fringe spacing variation is determined by the characteristic of the Gaussian beams. The shift of the fringe spacing distribution in the experiment is due to a slight tilt between the camera's surface and the optical axis. In order to investigate the accuracy of the model and the experiment, the relative differences between the modeling and experimental results at the same coordinates are estimated along the y- and z-axes, respectively, cf. Figure 7.

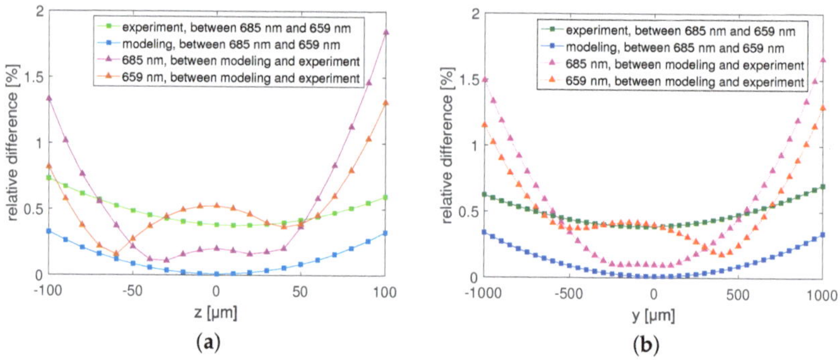

Figure 7. Relative differences between the modeling and experimental results on the z-axis (**a**) and the y-axis (**b**).

Figure 7a shows on the z-axis that the modeling results of both wavelengths offer tiny differences of below 0.33%, and for the experimental results they are below 0.74%. The relative differences between the modeling and experimental results with the same wavelength are all below 1.85%. Similarly, as depicted in Figure 7b, the modeling results of the two wavelengths gives the differences of below 0.34% on the y-axis. For the experimental results, their upper limit is 0.71%. With the same wavelength, the relative differences between the modeling and experimental results are limited to less than 1.67%. Eventually, the average relative differences are evaluated and listed in Table 1.

Table 1. Average relative differences in the numerical modeling and experiments with the two wavelengths.

	Average Relative Difference
Modeling of both wavelengths	0.12%
Experiments of both wavelengths	0.49%
Modeling and experiment of 659 nm	0.53%
Modeling and experiment of 685 nm	0.57%

This reveals that the average relative differences of the modeling and experimental results with wavelengths of 659 nm and 685 nm are all below 0.6% and, thus, indicates that the method is feasible.

Figure 7 also shows that the relative differences in the experiment are a little higher than in the modeling. This is caused by the noise of the fringe images and the slight position difference of the camera for the two fringe volumes. For the modeling, the fringe spacing difference results from the difference in wavelength. In the experiment, the trend of relative difference is influenced by the light intensity as well. The decreasing light intensity results in lowering the signal-to-noise ratio (SNR) approaching the edge of the beam. Therefore, the difference between the modeling and the experiment has a relatively significant increase at both ends of the curves. The minor variation of the differences around the center of the curve is due to a position deviation of the experiment relative to the model. Overall, the tendencies of the relative differences are comparable.

4. Conclusions

A full-field 3D mathematical model and experimental investigation of the interference fringe geometry in the intersection of Gaussian beams are proposed for minimizing the systematic error of laser Doppler sensors. The model is derived from the phase expression of Gaussian beams by using the theory of partial differentiation and introducing extension ratio, thus, can be universally applied to spherical and line-shaped Gaussian beams with arbitrary beam waist sizes and positions.

Based on a laser Doppler sensor system, the experimental fringe spacing analysis is conducted with line-shaped beams in the two wavelengths of 659 nm and 685 nm. A matrix camera integrated on a high-precision motion stage is used to detect fringe patterns at various positions inside of the intersection volume. The fringe spacings are then evaluated by employing the Fast Fourier Transform (FFT) along the fringe direction. Utilizing the same conditions, the modeling results show the average relative differences of 0.53% and 0.57% compared to the experimental results with both wavelengths, respectively. Between both wavelengths, the average relative differences in the modeling and experimental results are 0.12% and 0.49%, respectively. These validate the feasibility of the proposed methods.

As an outlook, the fringe geometry model can be further improved for arbitrary shaped Gaussian beams. The model refers to system parameters including focal length, wavelength, and grating constant for the dual-beam laser Doppler sensor configuration. It can offer a key priori knowledge of the appropriate parameters targeting different measurements that require a diverse measurement range, working distance, fringe spacing, etc. The model may also offer system design equations that identify the important parameters governing the fringe field uniformity. The limitation of systematic error is predictable by the model based on the modeled fringe geometry variation. Combined with fringe calibrations it allows for identifying alignment errors of the system for rapid system optimization. As a broad prospect, the model can be commonly applied to dual-beam interference sensors which are influenced by the nonuniform fringe field.

Author Contributions: Conceptualization, H.Z.; methodology, H.Z.; software, H.Z. and J.W.; validation, J.W. and S.W.; formal analysis, H.Z., J.W. and S.W.; investigation, H.Z. and J.W.; resources, H.Z.; data curation, J.W.; writing—original draft preparation, H.Z.; writing—review and editing, H.Z.; visualization, H.Z. and J.W.; supervision, H.Z.; project administration, H.Z.; funding acquisition, H.Z. All authors have read and agreed to the published version of the manuscript.

Funding: This research is funded by the National Natural Science Foundation of China (grant number 52205555) and the Fundamental Research Funds for the Central Universities (grant number 3132023106).

Institutional Review Board Statement: Not applicable.

Informed Consent Statement: Not applicable.

Data Availability Statement: The data underlying the results presented in this study are not currently publicly available but may be obtained from the authors upon request.

Conflicts of Interest: The authors declare no conflict of interest.

Appendix A

The mathematical proof from Equations (6) to (7) by differentiating with respect to x is given as

$$\begin{aligned}\frac{dn}{dx} =& -\frac{1}{\lambda}(-\sin\alpha - \sin\alpha) + \frac{1}{2\pi}\left[\frac{-z_{R_1}\sin\alpha}{z_1^2 + z_{R_1}^2} - \frac{z_{R_2}\sin\alpha}{z_2^2 + z_{R_2}^2}\right] \\ & - \frac{1}{2\lambda}\left[\frac{2x_1 z_1\cos\alpha(z_1^2 + z_{R_1}^2) + (x_1^2 + y_1^2)(z_1^2 - z_{R_1}^2)\sin\alpha}{(z_1^2 + z_{R_1}^2)^2}\right. \\ & \left. - \frac{2x_2 z_2\cos\alpha(z_2^2 + z_{R_2}^2) - (x_2^2 + y_2^2)(z_2^2 - z_{R_2}^2)\sin\alpha}{(z_2^2 + z_{R_2}^2)^2}\right],\end{aligned} \tag{A1}$$

$$
\begin{aligned}
\frac{dn}{dx} = & \frac{2\sin\alpha}{\lambda} + \frac{\sin\alpha}{2\pi}\left[\frac{-z_{R_1}}{z_1^2+z_{R_1}^2} - \frac{z_{R_2}}{z_2^2+z_{R_2}^2}\right] \\
& - \frac{\sin\alpha}{2\lambda}\left[\frac{2x_1z_1(z_1^2+z_{R_1}^2)\cdot(\tan\alpha)^{-1}+(x_1^2+y_1^2)(z_1^2-z_{R_1}^2)}{(z_1^2+z_{R_1}^2)^2}\right. \\
& \left. - \frac{2x_2z_2(z_2^2+z_{R_2}^2)\cdot(\tan\alpha)^{-1}+(x_2^2+y_2^2)(z_2^2-z_{R_2}^2)}{(z_2^2+z_{R_2}^2)^2}\right],
\end{aligned} \tag{A2}
$$

$$
\begin{aligned}
\frac{dn}{dx} = & \frac{2\sin\alpha}{\lambda}\left[1+\frac{\lambda}{4\pi}\left(\frac{-z_{R_1}}{z_1^2+z_{R_1}^2} - \frac{z_1}{z_2^2+z_{R_2}^2}\right)\right. \\
& - \frac{1}{4}\left(\frac{2}{\tan\alpha}\cdot\frac{x_1z_1}{z_1^2+z_{R_1}^2} + x_1^2\frac{z_1^2-z_{R_1}^2}{(z_1^2+z_{R_1}^2)^2} + y_1^2\frac{z_1^2-z_{R_1}^2}{(z_1^2+z_{R_1}^2)^2}\right. \\
& \left.\left. - \frac{2}{\tan\alpha}\cdot\frac{x_2z_2}{z_2^2+z_{R_2}^2} + x_2^2\frac{z_2^2-z_{R_2}^2}{(z_2^2+z_{R_2}^2)^2} + y_2^2\frac{z_2^2-z_{R_2}^2}{(z_2^2+z_{R_2}^2)^2}\right)\right].
\end{aligned} \tag{A3}
$$

Equation (7) is then achieved by taking the reciprocal of Equation (A3).

References

1. Kuschmierz, R.; Davids, A.; Metschke, S.; Löffler, F.; Bosse, H.; Czarske, J.; Fischer, A. Optical, in situ, three-dimensional, absolute shape measurements in CNC metal working lathes. *Int. J. Adv. Manuf. Technol.* **2016**, *84*, 2739–2749. [CrossRef]
2. Zhang, H.; Kuschmierz, R.; Czarske, J. Miniaturized interferometric 3-D shape sensor using coherent fiber bundles. *Opt. Lasers Eng.* **2018**, *107*, 364–369. [CrossRef]
3. Zhang, H.; Anders, D.; Löser, M.; Ihlenfeldt, S.; Czarske, J.; Kuschmierz, R. Non-contact, bi-directional tool tip vibration measurement in CNC milling machines with a single optical sensor. *Mech. Syst. Signal Process.* **2020**, *139*, 106647. [CrossRef]
4. Logozzo, S.; Valigi, M.C.; Canella, G. Advances in optomechatronics: An automated tilt-rotational 3D scanner for high-quality reconstructions. *Photonics* **2018**, *5*, 42. [CrossRef]
5. Dorsch, R.G.; Häusler, G.; Herrmann, J.M. Laser triangulation: Fundamental uncertainty in distance measurement. *Appl. Opt.* **1994**, *33*, 1306–1314. [CrossRef] [PubMed]
6. Norgia, M.; Pesatori, A. Interferometric instrument for thickness measurement on blown films. *Photonics* **2021**, *8*, 245. [CrossRef]
7. Yang, L.; Steinchen, W.; Schuth, M.; Kupfer, G. Precision measurement and nondestructive testing by means of digital phase shifting speckle pattern and speckle pattern shearing interferometry. *Measurement* **1995**, *16*, 149–160. [CrossRef]
8. Kempe, A.; Schlamp, S.; Rösgen, T.; Haffner, K. Low-coherence interferometric tip-clearance probe. *Opt. Lett.* **2003**, *28*, 1323–1325. [CrossRef] [PubMed]
9. Asundi, A.; Singh, V.R. Time-averaged in-line digital holographic interferometry for vibration analysis. *Appl. Opt.* **2006**, *45*, 2391–2395. [CrossRef]
10. Wang, Y.; Jiang, J.; Wanintrudal, C.; Du, C.; Zhou, D.; Smith, L.; Yang, L. Whole field sheet-metal tensile test using digital image correlation. *Exp. Tech.* **2010**, *34*, 54–59. [CrossRef]
11. Tao, X.; Fernandez, B.; Azucena, O.; Fu, M.; Garcia, D.; Zuo, Y.; Chen, D.C.; Kubby, J. Adaptive optics confocal microscopy using direct wavefront sensing. *Opt. Lett.* **2011**, *36*, 1062–1064. [CrossRef] [PubMed]
12. Zhang, H.; Kuschmierz, R.; Czarske, J.; Fischer, A. Camera-based speckle noise reduction for 3-D absolute shape measurements. *Opt. Express* **2016**, *24*, 12130–12141. [CrossRef]
13. Schawlow, A.L.; Townes, C.H. Infrared and optical masers. *Phys. Rev.* **1958**, *112*, 1940. [CrossRef]
14. Maiman, T.H. *Stimulated Optical Radiation in Ruby*; MacMillan & Company: New York, NY, USA, 1960.
15. Kogelnik, H.; Li, T. Laser beams and resonators. *Appl. Opt.* **1966**, *5*, 1550–1567. [CrossRef] [PubMed]
16. Tsang, W. A graded-index waveguide separate-confinement laser with very low threshold and a narrow Gaussian beam. *Appl. Phys. Lett.* **1981**, *39*, 134–137. [CrossRef]
17. Fan, R.; Liu, Z.; Jin, D.; Luo, T.; Li, N.; Li, S.; Wang, Y.; Xia, Y.; Lu, Z. High temporal waveform fidelity stimulated Brillouin scattering phase conjugate mirror using Novec-7500. *Opt. Express* **2023**, *31*, 1878–1887. [CrossRef]
18. Welsh, L.P.; Tuchman, J.A.; Herman, I.P. The importance of thermal stresses and strains induced in laser processing with focused Gaussian beams. *J. Appl. Phys.* **1988**, *64*, 6274–6286. [CrossRef]
19. Sun, J.; Wu, J.; Wu, S.; Goswami, R.; Girardo, S.; Cao, L.; Guck, J.; Koukourakis, N.; Czarske, J.W. Quantitative phase imaging through an ultra-thin lensless fiber endoscope. *Light. Sci. Appl.* **2022**, *11*, 204. [CrossRef]
20. Hanson, S. Broadening of the measured frequency spectrum in a differential laser anemometer due to interference plane gradients. *J. Phys. D Appl. Phys.* **1973**, *6*, 164. [CrossRef]

21. Durst, F.; Stevenson, W.H. Influence of Gaussian beam properties on laser Doppler signals. *Appl. Opt.* **1979**, *18*, 516–524. [CrossRef]
22. Durst, F.; Müller, R.; Naqwi, A. Measurement accuracy of semiconductor LDA systems. *Exp. Fluids* **1990**, *10*, 125–137. [CrossRef]
23. Miles, P.C. Geometry of the fringe field formed in the intersection of two Gaussian beams. *Appl. Opt.* **1996**, *35*, 5887–5895. [CrossRef]
24. Zhang, H. Laser interference 3-D sensor with line-shaped beam based multipoint measurements using cylindrical lens. *Opt. Lasers Eng.* **2022**, *159*, 107218. [CrossRef]
25. Svelto, O.; Hanna, D.C. *Principles of Lasers*; Springer: New York, NY, USA, 1998; Volume 4.
26. Truax, B.E.; Demarest, F.C.; Sommargren, G.E. Laser Doppler velocimeter for velocity and length measurements of moving surfaces. *Appl. Opt.* **1984**, *23*, 67. [CrossRef]
27. Matsubara, K.; Stork, W.; Wagner, A.; Drescher, J.; Müller-Glaser, K. Simultaneous measurement of the velocity and the displacement of the moving rough surface by a laser Doppler velocimeter. *Appl. Opt.* **1997**, *36*, 4516–4520. [CrossRef] [PubMed]

hν **photonics**

MDPI

Communication

Prediction of Shock Wave Velocity Induced by a Combined Millisecond and Nanosecond Laser Based on Convolution Neural Network

Jingyi Li [1,*], Wei Zhang [1], Ye Li [2] and Guangyong Jin [1]

[1] Jilin Key Laboratory of Solid-State Laser Technology and Application, School of Physics, Changchun University of Science and Technology, Changchun 130022, China; sjd@mails.cust.edu.cn (W.Z.); jgycust@cust.edu.cn (G.J.)

[2] School of Physics, Changchun University of Science and Technology, Changchun 130022, China; liye@cust.edu.cn

* Correspondence: 2022800023@cust.edu.cn

Abstract: The variation of shock-wave velocity with time induced by a millisecond-nanosecond combined pulse laser (CPL) on silicon is investigated. The convolution neural network (CNN) is used to predict the shock-wave velocity induced by a single ns laser and CPL with a ns laser energy density of 6, 12 and 24 J/cm^2, ms laser energy density of 0 and 226.13 J/cm^2, and pulse delay of 0, 0.4 and 0.8 ms. The four-layer CNN model was applied, ns laser energy density, ms laser energy density, pulse delay and time were set as the input parameter, while the shock-wave velocity was set as the output parameter. The correlation coefficient (R^2), mean absolute error (MAE) and root mean square error ($RMSE$) of the CNN model on the test data set was 0.9865, 3.54 and 3.01, respectively. This indicated that the CNN model shows a high reliability in the prediction of CPL-induced shock-wave velocity with limited experimental data.

Keywords: convolution neural network; shock wave velocity; combined pulse laser

1. Introduction

The aerospace field is developing extremely fast nowadays, and space debris are gradually increasing. The traditional method of space debris removal is mechanical capture, through which it is difficult to remove small space debris [1–3]. The shock wave generated on the target debris that are irradiated by a high-energy pulsed laser can effectively remove small space debris. When the target is irradiated by a pulsed laser, plasma is generated on the surface. After the laser stops irradiating, a shock wave is driven by extending plasma continuing to propagate [4–6]. From a macro perspective, this produces a certain recoil pressure on the target. Scholars from all over the world have mainly studied the short-pulse-laser-induced shock-wave pressure field [7–10]. For example, J. Radziejewskaa et al. studied the velocity and pressure of a shock wave produced by a nanosecond pulse laser with a wavelength of 1064 nm and pulse width of 12 ns, and they found a qualitative correlation between the shock-wave velocity and pressure: they found that the greater the shock-wave velocity, the greater the pressure [11]. Daniel J. and Frster D J et al. used a picosecond pulse laser with a 532 nm wavelength and 10 ps pulse width to irradiate copper. The thrust produced by single, double and triple pulse lasers was studied. It is found that the mass specific thrust produced by a double pulse laser is almost three times that of a single pulse laser. However, the mass specific thrust of a triple pulse laser is basically the same as that of a single pulse laser. This is mainly due to the ablation effect of the third pulse laser being obviously stronger than that of the double pulse laser [12]. Kiran P. P. et al. used two-dimensional emission and shadow imaging techniques to study the interaction between two plasmas and shock waves induced by nanosecond laser pulses with a wavelength of 532 nm and pulse width of 7 ns. The effects of the distance between the two plasma sources and their energy ratio on the evolution of the plasma and shock wave were provided. It was found that the shock wave produced by the high-energy plasma source propagated through the lower energy plasma and led to the generation of a plasma jet. With the increase of the energy ratio, the diameter of the plasma jet will increase [13].

With the increasing demand for propulsion efficiency through the use of shock-wave-induced recoil pressure, the method for obtaining shock-wave acceleration is very significant. However, the laser-crystal material damage threshold makes it impossible to increase laser energy infinitely to obtain an increment of shock-wave velocity. In recent years, scholars have gradually converted from single pulse lasers to CPLs to irradiate a target and produce plasma and shock waves [14–18]. The ms-ns CPL can not only improve the laser drilling efficiency [19–21], but it also induces the phenomenon of shock-wave acceleration [22]. For example, Yuan B.S. and Wang D. et al. used a combination of a millisecond laser and nanosecond laser with a wavelength of 1064 nm and pulse width of 1 ms and 10 ns, respectively, to irradiate aluminum alloy. It was found that the ablation depth produced by the combined pulse laser was nine times that of the single pulse laser [21]. Li J. and Zhang W. et al. used a combined pulse laser with a millisecond pulse laser and nanosecond pulse laser with different energy ratios and pulse delays to irradiate monocrystalline silicon, and the variation in the shock-wave velocity was studied. It was found that under the conditions of an appropriate energy ratio and pulse delay, the shock-wave velocity produced by the CPL was 1.1 times higher than that of the single nanosecond pulse laser [22]. Due to the large amount of CPL parameters, different parameters of CPL will lead to a change in shock-wave velocity, and it has nonlinear variation characteristics [23]. The existing shock-wave-velocity measurement methods are mainly off-line measurements. This is inefficient and makes it difficult to acquire the CPL parameters of the highest shock-wave velocity.

Nowadays, artificial intelligence deep-learning is developing rapidly. Scholars are committed to solving nonlinear problems in the laser-material interaction field, mainly focusing on the prediction of target properties after laser shock peening [24,25]. For instance, Jiajun Wu et al. uses an artificial neural network to predict the mechanical properties of titanium alloy after laser shock peening. The results show that the correlation index R^2 of the test data set is 0.997 and 0.987 [26]. As for the prediction of laser-induced shock-wave velocity, it mainly focuses on the prediction of shock-wave velocity induced by a single pulse laser. For example, Matsui K et al. used the least square method to predict a laser-supported detonation wave velocity. It is indicated that the greater the laser energy density, the greater the laser-supported detonation wave velocity [27]. However, there is a nonlinear relationship between the multi-dimensional parameters of CPL and the shock-wave velocity. Moreover, there are few reports on the application of CNN to predicting CPL-induced shock-wave velocity.

With the aim of acquiring the laser parameters of the highest shock-wave velocity, the variation law of CPL-induced shock-wave velocity on silicon is studied. The CNN algorithm is used to predict the ns-laser- and CPL-induced shock-wave velocity. The ns laser energy density, the ms laser energy density, pulse delay and time are set as input parameters, and the shock-wave velocity is set as the output parameter. The mean impact value algorithm (MIV) is used to analyze the ranking importance of the input parameters' effect on the shock-wave velocity. This work provides important guidance for using appropriate methods to predict shock-wave velocity under conditions of limited data. Moreover, it can also provide theoretical support for the application of laser space debris removal technology, and it is of great significance in ensuring the safe development of space activities.

2. Material and Method

2.1. Experiment

The laser-induced shock-wave experiment was performed on a combined pulse laser system, which consisted of a nanosecond pulse laser (ns laser) and a millisecond pulse laser (ms laser). The laser parameters are listed in Table 1; the ms laser beam and ns laser beam pass through lenses L1 ($f = 500$ mm) and L2 ($f = 500$ mm), respectively, and irradiate onto the same point of the target at a space angle of 5 degrees. In this paper, the shock-wave propagation process is monitored by the optical shadow method. The principle of the optical shadow method is to transform the density gradient change in the flow field into the relative light intensity. A 532 nm continuous laser is used as the background light source in order to be vertically incident to the generation area of the shock wave in front of the target through the beam expander. After passing lens L3 and attenuator film F1, the shock-wave image is recorded by a high-speed camera. The resolution of the high-speed camera is 384 × 160, the frame rate is 200,000 fps, and the exposure time is 1/6,300,000 s. The millisecond pulse laser energy was recorded in real time by using a Spectroscope S1 and an energy meter. The pulse delay of the two lasers is controlled by DG645 and monitored by an oscilloscope and two photodetectors PD1 and PD2 in real time. We defined the time interval when a nanosecond pulse laser lags behind a ms laser. The laser-induced shock-wave experiment setup in this work is shown in Figure 1.

Table 1. The laser parameters.

Laser Parameters	Value
Laser wavelength	1064 nm
Pulse width for ms laser	1 ms
Pulse width for ns laser	10 ns
Spot diameter for ms laser	1.3 mm
Spot diameter for ns laser	1 mm
Energy density of ms laser	226.13 J/cm^2
Energy density of ns laser	6, 12 and 24 J/cm^2
Repetition rate	1 Hz

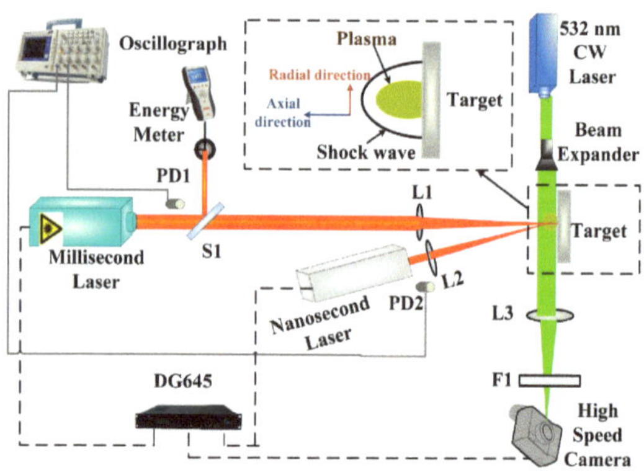

Figure 1. The CPL-induced shock-wave experimental setup.

The investigated target is N-doped (100) monocrystalline silicon, and the dimension of the experimental sample is 4 mm × 12.7 mm (thickness × radius). The experiment was carried out at 20 °C, under normal atmospheric pressure, in air. The shock wave velocity can be calculated from Equation (1):

$$v(t) = \frac{S(t+\tau) - S(t)}{\tau}, \quad (1)$$

where $S(t)$ is the maximum axial propagation distance of the shock wave at time t. τ is the time interval between the above two measurements.

2.2. CNN Model

The effects of the ms-ns CPL parameters, including the ns laser energy density, ms laser energy density, and pulse delay, on the shock-wave velocity were investigated. The CNN model is used to predict the CPL-induced shock-wave velocity. The CNN model structure is shown in Figure 2. The ns laser energy density, the ms laser energy density, pulse delay and time were set as the input, while the shock-wave velocity was set as the output. The experimental data induced by ms-ns CPL with $D_{ns} = 6\ J/cm^2$ and $24\ J/cm^2$ are selected as training sets, and the experimental data induced by ms-ns CPL with $D_{ns} = 12\ J/cm^2$ are set as testing sets. The input conditions are shown in Table 2. As a result, according to the characteristics of the data, the convolution layer is set to one layer. In order to ensure the data integrity of the shock-wave velocity, the pool layer is set to be removed. To avoid the loss of data feature in the convolution layer, set the edge zeroing operation. In order to comprehensively consider the training effect during the training process, the number of epochs is 2500, and the learning rate is 0.01. The specific convolution operation is shown in Formula (2):

$$x = \sum_{i=1}^{c} w_{i,c} * x_i + b_i, \quad (2)$$

where x_i is the output of the ith channel in the input layer, c is the convolution layer channel, x is the convolution layer output, $w_{i,c}$ is the weight matrix of the convolutional kernel in the convolution layer, b_i is the bias term, and $*$ is the convolution operation.

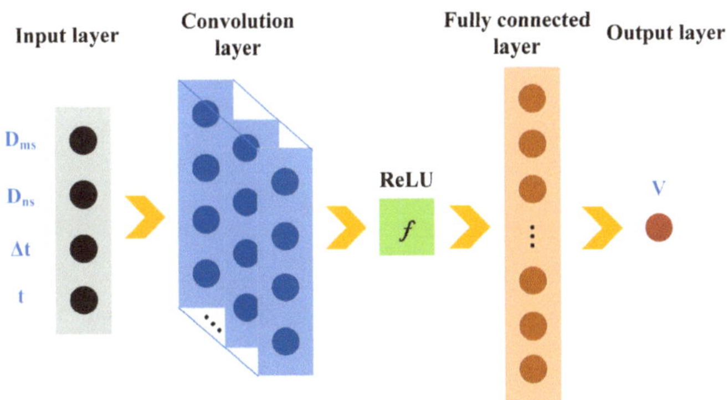

Figure 2. Schematic of CNN structure.

Table 2. The input conditions for CNN.

Condition Number	ns Laser Energy Density D_{ns} (J/cm^2)	ms Laser Energy Density D_{ms} (J/cm^2)	Pulse Delay Δt (ms)
1	6	0	0
2	6	226.13	0.4
3	6	226.13	0.8
4	12	0	0
5	12	226.13	0.4
6	12	226.13	0.8
7	24	0	0
8	24	226.13	0.4
9	24	226.13	0.8

A batch normalization operation is added after the convolution layer to improve the performance and enhance the generalization ability of the network. The specific calculation process is shown in Formula (3):

$$x' = \gamma\left(\frac{x - u_B}{\sqrt{\sigma_B^2 + \varepsilon}}\right) + \beta, \tag{3}$$

where x' is the batch normalization result of the convolutional kernel output value. u_B is the average of the input value. σ_B^2 is the standard deviation of the input value. ε is a constant vector. γ and β denote scale factors and shift factors, respectively. Then, the activation function is used to transform its output characteristics nonlinearly and accelerate the model convergence, and the calculation process is shown in Formula (4):

$$RELU(x') = \max(0, x'), \tag{4}$$

the full connection layer converts the features into one-dimensional feature vectors and extracts the features again, before finally outputting a predicted value. The operation is shown in Formula (5):

$$y = RELU\left((w)^T x' + b\right), \tag{5}$$

where w is the weight, y is the output of the fully connected layer, and b is the offset term. The CNN model performance was evaluated by the R^2, $RMSE$ and MAE. These indices are defined as follows [28,29]:

$$R^2 = 1 - \frac{\sum_{i=1}^{n}(y_i - \hat{y}_i)^2}{\sum_{i=1}^{n}(y_i - \overline{y})^2}, \tag{6}$$

$$RMSE = \sqrt{\frac{1}{n}\sum_{i=1}^{n}(y_i - \hat{y}_i)^2}, \tag{7}$$

$$MAE = \frac{1}{n}\sum_{i=1}^{n}|y_i - \hat{y}_i|, \tag{8}$$

where n is the number of predicted values, y represents the original experimental value, $\hat{y}_i$ defines the CNN predicted value, and $\overline{y}$ stands for the experimental mean value.

3. Results and Discussion

3.1. Experimental Shock-Wave Velocity

The shock-wave propagation morphology induced by the ns laser is shown in Figure 3a. The plasma plume is generated on the surface at 2 μs, the plasma stops expanding, and the shock wave propagates at 7 μs. The shock-wave propagation morphology induced by the ms laser is shown in Figure 3b. Compared with Figure 3a, the ms laser produces thin plasma on the surface, and the plasma expands very slowly. This is mainly because the ms laser belongs to the long pulse laser, and its peak power density is lower than that of the ns laser. In order to analyze the propagation morphology and velocity of the shock wave induced by CPL under different pulse delay conditions, the time when the ns laser begins to irradiate the target is defined as the time starting point. The shock-wave propagation morphology induced by the CPL with different pulse delays is shown in Figure 3c,d. It can be found that the ms laser first produces thin plasma on the target, after which a large amount of plasma overflows the surface while the ns laser irradiates the target. This is mainly due to the thermal accumulation effect of ms-laser-induced plasma, while the ns laser is mainly due to the ionization effect.

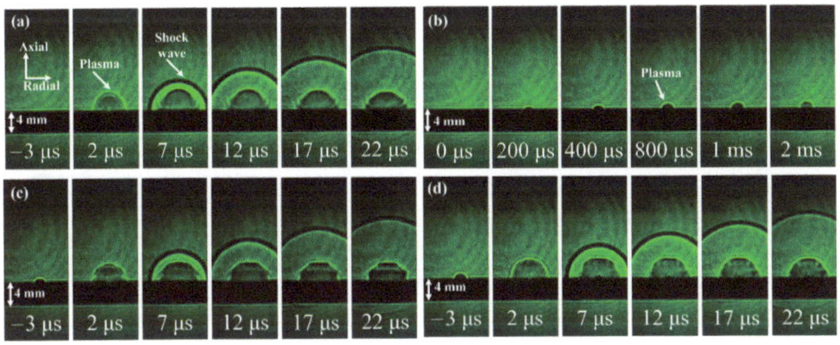

Figure 3. The shock-wave propagation morphology induced by (**a**) ns laser, (**b**) ms laser and ms-ns CPL with pulse delays of (**c**) 0.4 ms and (**d**) 0.8 ms, with $D_{ns} = 12\ J/cm^2$, $D_{ms} = 226.13\ J/cm^2$.

Figure 4 displays the experimental shock-wave velocity of the ns laser and CPL varying with time. The velocity of the shock wave decreases with time. The shock-wave velocity induced by ns laser with an energy density of 6 J/cm^2 is the lowest among the above experimental conditions; the velocity reaches 391.8 m/s at 7 μs and decreases to 338 m/s at 32 μs. Under the condition of $D_{ns} = 6\ J/cm^2$, the shock-wave velocity induced by the CPL with a pulse delay of 0.8 ms reaches the maximum. However, the shock-wave velocities induced by the CPL with pulse delays of 0.4 ms and 0.8 ms are close at 7 μs, and the difference in shock-wave velocities increases after 12 μs. Under the condition of $D_{ns} = 12\ J/cm^2$, the shock-wave velocity induced by CPL with a pulse delay of 0.4 ms is the highest. The velocity reaches 457 m/s at 7 μs and decreases to 360 m/s at 32 μs. The shock-wave velocity induced by CPL with $D_{ns} = 24\ J/cm^2$, $D_{ms} = 226.13\ J/cm^2$, and a pulse delay of 0.4 ms is the highest; the velocity reaches 512 m/s at 7 μs and decreases to 368 m/s at 32 μs. We found that with the increase of the ns laser energy density, the pulse delay of the maximum shock-wave velocity changes from 0.8 ms to 0.4 ms. When the energy density of the ns laser is 6 J/cm^2 and 12 J/cm^2, the thermal ionization mechanism plays a leading role. The plasma induced by the ms laser does not absorb ns laser energy but provides initial free electrons for the plasma induced by the ns laser; the collision ionization probability of free electrons and silicon atoms is increased. Thus, the shock-wave velocity increases. Therefore, when the pulse delay is 0.8 ms, the shock-wave velocity reaches the maximum. When the energy density of the ns laser increases to 24 J/cm^2, the ionization ability of the ns laser to the target becomes stronger. With the increase of the pulse delay, the plasma density induced by the ms laser increases. This will produce a combination effect of thermal ionization mechanism and reverse toughening absorption mechanism; that is, the ns laser energy is absorbed by both the plasma induced by the ms laser and the target. Moreover, the higher the plasma density induced by the ms laser, the more obvious the reverse toughening absorption mechanism. This leads to the decrease of the plasma expansion velocity induced by the thermal ionization mechanism of the ns laser. Therefore, the shock-wave velocity reaches the maximum value at 0.4 ms.

Figure 5 shows the CPL-induced shock-wave-velocity increment under the condition of different pulse delays and a ns laser energy density of 7 μs. The velocity increment is the difference between the CPL-induced shock-wave velocity and the ns-laser-induced shock-wave velocity. Compared with the shock-wave velocity induced by the ns laser, the CPL can induce shock-wave acceleration. This is mainly due to the fact that the ms-laser-induced plasma provides the initial electrons for the ns-laser-induced plasma, the density of free electrons increases, the probability of collision ionization between free electrons and silicon atoms increases, and the expansion velocity of ns-laser-induced plasma increases. Finally, this leads to the acceleration phenomenon of a shock wave. When $D_{ns} = 6\ J/cm^2$, the CPL-induced shock-wave velocity increment reaches the maximum with pulse delays of 0.4 ms and 0.8 ms. However, when $D_{ns} = 24\ J/cm^2$, the velocity increment of the shock wave induced by CPL is at a minimum, with pulse delays of 0.4 ms and 0.8 ms. Therefore, the larger the energy density of the ns laser, the smaller the CPL-induced shock-wave-velocity increment. This is mainly due to the fact that with the increase of ns laser energy, the plasma density and the number of steam atoms increase, while the free electrons in the plasma induced by the ms laser are limited. Therefore, the acceleration efficiency of collision ionization between free electrons and steam atoms decreases.

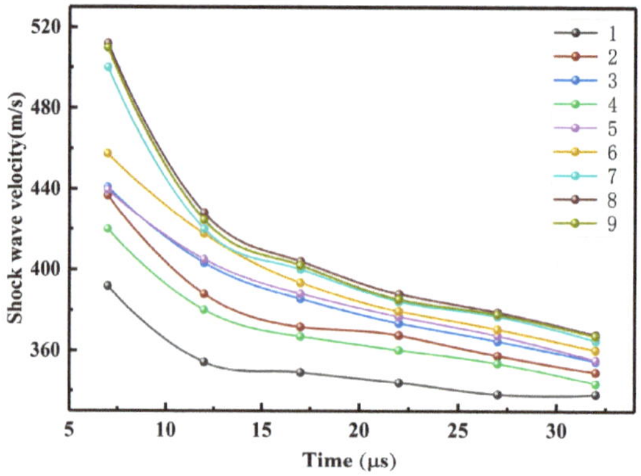

Figure 4. The variation trend of shock-wave velocity with time.

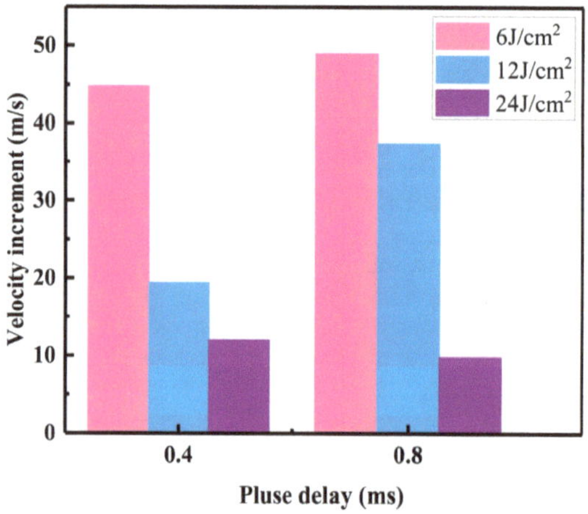

Figure 5. The shock-wave-velocity increment of the CPL relative to the ns laser.

3.2. CNN Prediction for Shock-Wave Velocity

The experimental shock-wave velocity with Conditions 1 to 3 and Conditions 7 to 9 is set to the training set, and the experimental shock-wave velocity with Conditions 4 to 6 is set to the testing set. Figure 6a,b illustrate the comparison between the experimental and predicted shock-wave velocity, and the shock-wave velocity predicted by the CNN model corresponds well with the experimental shock-wave velocity. The CNN model can learn the shock-wave propagation velocity induced by CPL well and predict the variation of the shock-wave velocity with time under different conditions well. Therefore, the CNN model can accurately predict the shock-wave velocity induced by CPL in the case of limited data. The corrections of the training data and test data are shown in Figure 6c,d. The scattered points are all distributed near the guideline of the training and testing sets. The results show that the CNN model correctly predicts the shock-wave velocity induced by CPL.

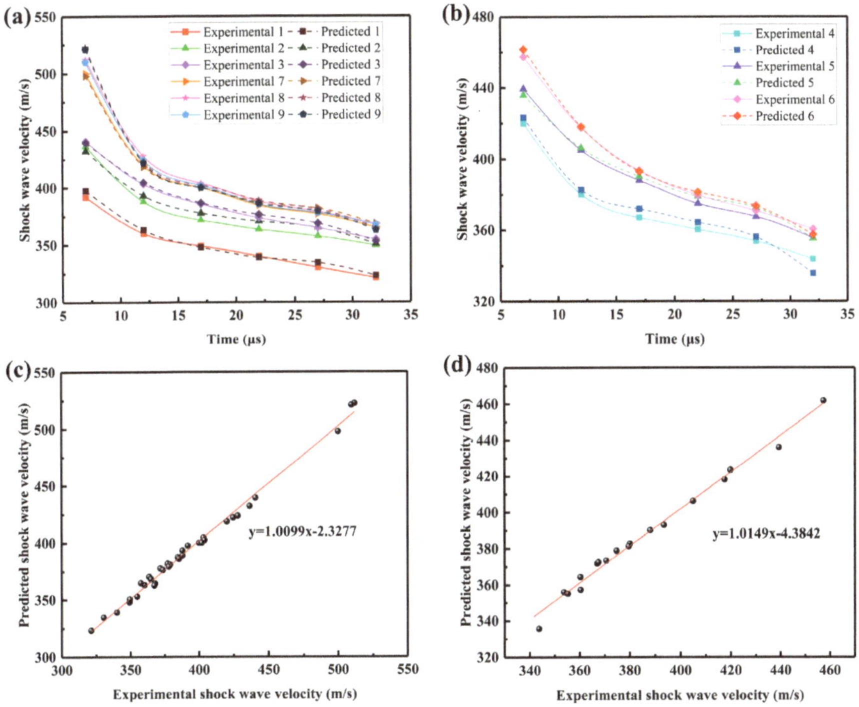

Figure 6. The CNN model results: (**a**) Comparison between experiment and prediction of shock-wave velocity on training data set, (**b**) comparison between experiment and prediction of shock-wave velocity on test data set, (**c**) correlation of the training data set, and (**d**) correlation of the test data set.

The CNN model's performance parameters on the training set and test set are listed in Table 3. The performance of the test set is the key factor to assess the CNN model's performance. The R^2 of the test data set is 0.9865, which is smaller than that of the training data set (0.9912), but it remains in great accordance with the experimental and predicted shock-wave velocity. The *RMSE* and *MAE* of the test data set are 3.54 and 3.01, respectively. The value of *RMSE* and *MAE* is relatively low, and the CNN model can accurately predict the shock-wave velocity. Therefore, the CNN can be used in the prediction of CPL-induced shock-wave velocity.

Table 3. The CNN model's performance parameters.

Training Data			Test Data		
R^2	*RMSE*	*MAE*	R^2	*RMSE*	*MAE*
0.9912	4.18	3.27	0.9865	3.54	3.01

In addition to the prediction of the shock-wave velocity, we also use the MIV algorithm to obtain the importance of network input parameters to the predicted value, which is shown in Figure 7. The ranking importance for predicting laser shock-wave velocity is ns laser energy > time > ms laser energy > pulse delay. Due to the fact that the greater the velocity of the shock wave, the greater the pressure on the surface, in order to obtain a higher shock-wave velocity, we should give priority to ns laser energy, followed by ms laser energy, and finally pulse delay.

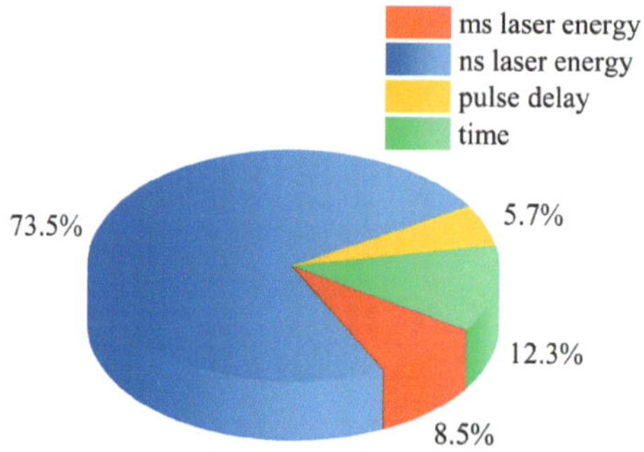

Figure 7. Ranking importance of input parameters.

4. Conclusions

In this paper, the variation of shock-wave velocity with time via ms-ns CPL is studied, and the prediction model of shock-wave velocity is established based on CNN. The ns laser energy density, ms laser energy density, pulse delay and time are input parameters, while the shock-wave velocity is the output parameter. Finally, the MIV algorithm is used to analyze the importance of network input parameters to the predicted value. The following are the primary conclusions:

(1) Compared with the single ns laser, the CPL can induce shock-wave acceleration. Furthermore, the CPL-induced shock-wave velocity increment is inversely proportional to the energy density of the ns laser.
(2) The predicted shock-wave velocity obtained by the CNN model fits well with the experimental value. The R^2, *RMSE* and *MAE* of the CNN model on the test data set are 0.9865, 3.54 and 3.01, respectively. It is indicated that the CNN model has a high reliability for predicting the shock-wave velocity.
(3) In order to obtain a higher shock-wave velocity, the ranking importance of the input parameters' effect on the shock-wave velocity is: ns laser energy > ms laser energy > pulse delay. This provides a theoretical basis when setting ms-ns CPL parameters with a higher shock-wave velocity.

Author Contributions: Writing—Review & Editing, J.L.; Methodology, W.Z.; Conceptualization, G.J. and Y.L. All authors have read and agreed to the published version of the manuscript.

Funding: This work was supported by the National Natural Science Foundation of China (Grant No. U19A2077), Jilin Provincial Education Department Project (No. JJKH20230796KJ) and Changchun Science and Technology Development Plan Project (No. 21ZY34).

Institutional Review Board Statement: Not applicable.

Informed Consent Statement: Not applicable.

Data Availability Statement: Not applicable.

Acknowledgments: We are grateful to the Jilin Key Laboratory of Solid-State Laser Technology and Application for providing us with a scientific research guarantee.

Conflicts of Interest: The authors declare no conflict of interest.

References

1. Phipps, C.R.; Baker, K.L.; Libby, S.B.; Liedahl, D.A.; Olivier, S.S.; Pleasance, L.D.; Rubenchik, A.; Trebes, J.E.; George, E.V.; Marcovici, B.; et al. Removing orbital debris with lasers. *Adv. Space Res.* **2012**, *49*, 1283–1300. [CrossRef]
2. Shen, S.; Jin, X.; Hao, C. Cleaning space debris with a space-based laser system. *Chin. J. Aeronaut.* **2014**, *27*, 805–811. [CrossRef]
3. Nishida, S.I.; Kawamoto, S.; Okawa, Y.; Terui, F.; Kitamura, S. Space debris removal system using a small satellite. *Acta Astronaut.* **2009**, *65*, 95–102. [CrossRef]

4. Pasley, J.; Bush, I.A.; Robinson, A.P.; Rajeev, P.P.; Mondal, S.; Lad, A.D.; Ahmed, S.; Narayanan, V.; Kumar, G.R.; Kingham, R.J. Generation of shock waves in dense plasmas by high-intensity laser pulses. *Nukleonika* **2015**, *60*, 193–198. [CrossRef]
5. Garbaruk, A.V.; Gritskevich, M.S.; Kalmykov, S.G.; Sasin, M.E. Computational simulation of laser plasma emission with shock-wave-affected density distribution in the gas-jet target. *Technol. Phys. Lett.* **2016**, *42*, 993–996. [CrossRef]
6. Ukai, T.; Kontis, K.; Yang, L. Flow structure generated by laser-induced blast wave propagation through the boundary layer of a flat plate. *Aerosp. Sci. Technol.* **2018**, *78*, 569–573. [CrossRef]
7. Yoh, J.J.; Lee, H.; Choi, J.; Lee, K.; Kim, K. Ablation-induced explosion of metal using a high-power Nd: YAG laser. *J. Appl. Phys.* **2008**, *103*, 043511. [CrossRef]
8. Gregorčič, P.; Možina, J. High-speed two-frame shadowgraphy for velocity measurements of laser-induced plasma and shockwave evolution. *Opt. Lett.* **2011**, *36*, 2782. [CrossRef]
9. Cao, S.; Su, M.; Min, Q.; Sun, D.; Ma, P.; Wang, K.; Jiao, Z.; Dong, C. Dynamics and density distribution of laser-produced Al plasmas using optical interferometry and optical emission spectroscopy. *J. Quant. Spectrosc. Radiat. Transf.* **2019**, *225*, 69–75. [CrossRef]
10. Kraft, S.; Schille, J.; Mauersberger, S.; Schneider, L.; Loeschner, U. Pump-probe imaging for process control and optimization in high-speed laser micro machining. In Proceedings of the Laser-based Micro- and Nanoprocessing XIV, San Francisco, CA, USA, 12 March 2020.
11. Radziejewska, J.; Strzelec, M.; Ostrowski, R.; Sarzyński, A. Experimental investigation of shock wave pressure induced by a ns laser pulse under varying confined regimes. *Opt. Laser. Eng.* **2020**, *126*, 105913. [CrossRef]
12. Frster, D.J.; Faas, S.; Weber, R.; Thomas, G. Thrust enhancement and propellant conservation for laser propulsion using ultra-short double pulses. *Appl. Surf. Sci.* **2020**, *510*, 145391. [CrossRef]
13. Guthikonda, N.; Manikanta, E.; Chelikani, L.; Shiva, S.S.; Harsha, S.S.; Ikkurthi, V.R.; Kiran, P.P. Interaction of two counterpropagating laser induced plasmas and shock waves in air. *Phy. Plasma.* **2020**, *27*, 023107. [CrossRef]
14. Cao, S.; Su, M.; Ma, P.; Wang, K.; Dong, C. Expansion dynamics and emission characteristics of nanosecond–picosecond collinear double pulse laser-induced Al plasma in air. *J. Quant. Spectrosc. Radiat. Transf.* **2019**, *242*, 106773. [CrossRef]
15. Cao, S.; Su, M.; Liu, J.; Min, Q.; Dong, C. Expansion dynamics and compression layer in collinear double-pulse laser produced plasmas in a vacuum. *Phys. Plasmas* **2020**, *27*, 052101. [CrossRef]
16. Smijesh, N.; Rao, K.H.; Chetty, D.; Litvinyuk, I.V.; Sang, R.T. Plasma plumes produced by laser ablation of Al with single and double pulse schemes. *Opt. Lett.* **2017**, *43*, 6081–6084. [CrossRef] [PubMed]
17. Wang, Q.; Qi, H.; Zeng, X.; Chen, A.; Gao, X.; Jin, M. Time-resolved spectroscopy of collinear femtosecond and nanosecond dual-pulse laser-induced Cu plasmas. *Plasma Sci. Technol.* **2021**, *23*, 121–127. [CrossRef]
18. Jia, X.; Chen, Y.; Zhu, G.; Wang, H.; Zhu, X. Experimental study on the optimum matching of CW-nanosecond combined pulse laser drilling. *Appl. Opt.* **2019**, *58*, 9105. [CrossRef]
19. Pan, Y.; Lv, X.; Zhang, H.; Chen, J.; Han, B.; Shen, Z.; Lu, J.; Ni, X. Millisecond laser machining of transparent materials assisted by a nanosecond laser with different delays. *Opt. Lett.* **2016**, *41*, 2807–2810. [CrossRef]
20. Yuan, B.; Zhang, Y.; Zhang, W.; Dong, Y.; Jin, G. The Effect of Spot Size Combination Mode on Ablation Morphology of Aluminum Alloy by Millisecond-Nanosecond Combined-Pulse Laser. *Materials* **2018**, *11*, 1419. [CrossRef]
21. Yuan, B.S.; Wang, D.; Dong, Y.; Zhang, W.; Jin, G.Y. Experimental study of the morphological evolution of the millisecond–nanosecond combined-pulse laser ablation of aluminum alloy. *Appl. Opt.* **2018**, *57*, 5743–5748. [CrossRef]
22. Li, J.; Zhang, W.; Zhou, Y.; Yuan, B.; Cai, J.; Jin, G. The acceleration mechanism of shock wave induced by millisecond-nanosecond combined-pulse laser on silicon. *Plasma Sci. Technol.* **2021**, *23*, 055507. [CrossRef]
23. Li, J.; Zhang, W.; Li, Y.; Jin, G. The Acceleration Phenomenon of Shock Wave Induced by Nanosecond Laser Irradiating Silicon Assisted by Millisecond Laser. *Photonics* **2023**, *10*, 260. [CrossRef]
24. Wu, J.; Li, Y.; Jibin, Z.; Qiao, H.; Yang, Y. Prediction of residual stress induced by laser shock processing based on artificial neural networks for FGH4095 superalloy. *Mater. Lett.* **2021**, *286*, 129269. [CrossRef]
25. Wu, J.; Xu, Z.; Qiao, H.; Zhao, J.; Huang, Z. Mechanical properties prediction of superalloy FGH4095 treated by laser shock processing based on machine learning. *Mater. Lett.* **2021**, *297*, 129970. [CrossRef]
26. Wu, J.; Huang, Z.; Qiao, H.; Zhao, Y.; Li, J.; Zhao, J. Artificial neural network approach for mechanical properties prediction of TC4 titanium alloy treated by laser shock processing. *Opt. Laser Technol.* **2021**, *143*, 107385. [CrossRef]
27. Matsui, K.; Shimano, T.; Ofosu, J.A.; Komurasaki, K.; Schoenherr, T.; Koizumi, H. Accurate propagation velocity measurement of laser supported detonation waves. *Vacuum* **2016**, *136*, 171–176. [CrossRef]
28. András, F.; György, C.; András, H. A Convolutional Neural Network with a Wave-Based Convolver. *Electronics* **2023**, *12*, 1126.
29. Xing, J.; Xu, J. An Improved Convolutional Neural Network for Recognition of Incipient Faults. *IEEE Sens. J.* **2022**, *22*, 16314–16322. [CrossRef]

hv photonics

MDPI

Article

Analytical Model of Point Spread Function under Defocused Degradation in Diffraction-Limited Systems: Confluent Hypergeometric Function

Feijun Song [1,†], Qiao Chen [1,†], Xiongxin Tang [1,2,*] and Fanjiang Xu [1,2]

1 Institute of Software, Chinese Academy of Sciences, Beijing 100190, China
2 University of Chinese Academy of Sciences, Beijing 100049, China
* Correspondence: xiongxin@iscas.ac.cn
† These authors contributed equally to this work.

Abstract: In recent years, optical systems near the diffraction limit have been widely used in high-end applications. Evidently, an analytical solution of the point spread function (PSF) will help to enhance both understanding and dealing with the imaging process. This paper analyzes the Fresnel diffraction of diffraction-limited optical systems in defocused conditions. For this work, an analytical solution of the defocused PSF was obtained using the series expansion of the confluent hypergeometric functions. The analytical expression of the defocused optical transfer function is also presented herein for comparison with the PSF. Additionally, some characteristic parameters for the PSF are provided, such as the equivalent bandwidth and the Strehl ratio. Comparing the PSF obtained using the fast Fourier transform algorithm of an optical system with known, detailed parameters to the analytical solution derived in this paper using only the typical parameters, the root mean square errors of the two methods were found to be less than 3% in the weak and medium defocus range. The attractive advantages of the universal model, which is independent of design details, objective types, and applications, are discussed.

Keywords: point spread function; diffraction-limited system; analytical model; confluent hypergeometric function

1. Introduction

The point spread function (PSF) is a term used to describe the intensity distribution of the image that is received at the detector plane when the light emitted by a point source passes through an optical system [1–3]. This function can be utilized to evaluate the performance of an optical system. The image captured by the optical system can be seen as the convolution of the actual object and the PSF. Thus, the PSF is often used as a physical reference in image processing and computer vision, particularly for image restoration, image denoising, and target detection [4–6]. Chen et al. [7,8] employed ray tracing to compute the PSF and utilized the spatial frequency response measurements to optimize the perturbation parameters of the optical system. They successfully developed a virtual image camera that produced imaging simulation results closely matching real-world photography. Furthermore, by integrating a deep learning algorithm, they were able to mitigate optical aberrations arising from processing errors. Researchers from the Chinese Academy of Sciences, Tsinghua University, and the Howard Hughes Medical Institute fused physical priors, such as the PSF of the optical system, with a neural network architecture design to achieve high-performance and high-fidelity microscopic image denoising and super-resolution reconstruction [9]. Additionally, Chan et al. from Carnegie Mellon University [10] have put forth a groundbreaking approach to compressive high-speed imaging. By time-encoding the camera's PSF, their method provides better optical efficiency and can be applied across a broader spectrum of scene categories. Yoav Shechtman et al. [11] proposed a framework for pupil-plane modulation in 3D imaging applications that require precise localization, based on the PSF of the system. This framework aims to extract maximum physical information about the position of a single nanoscale object from an image. The research group of Zhang Yunhai at the Suzhou Institute of Biomedical Engineering Technology, Chinese Academy of Sciences, has proposed a 2π-DH-PSF using the Fresnel zone approach that can rotate 2π radians, showcasing its superior performance in the 3D localization imaging of nanoparticles [12].

Diffraction-limited systems are those in which the diffraction dominates the image, and the geometric aberrations are of the same order compared to the Airy disk. With high-level design algorithms and the advancement of precision optical and mechanical machining capabilities, an increasing number of diffraction-limited optical systems are being designed and utilized in high-end technology fields such as aerospace, medicine, and astronomy [13–15]. Due to the more rigorous requirements regarding imaging quality within diffraction-limited optical systems, physical intelligence-driven image post-processing algorithms are frequently utilized to enhance image quality, in addition to focusing. The implementation of intelligent image-processing techniques, grounded in physical priors, hinges on the precise determination of the PSF for each defocusing position [16]. Therefore, there is an immediate need to investigate a high-precision, low-complexity solution for the PSF.

The most widely used method for calculating the PSF is via the Fourier transform of the wavefront using the fast Fourier transform algorithm, necessitating the modeling of the optical system to obtain the optical path difference through ray tracing [7,17–20]. While this method is commonly employed for PSF calculations in optical systems, its specific implementation remains challenging. Firstly, the ray tracing method necessitates precise knowledge of the optical system's detailed parameters, such as material properties, curvature, thickness, spacing, and other intricate lens details, which may be difficult to acquire for various reasons. Additionally, complex optical systems require significant computational resources for the process of ray tracing, which leads to a slow calculation process; analytical solutions can alleviate these problems to a certain extent. Various research papers have explored the analysis and analytical calculation of the PSF. Lommel first provided an analytical expression for the PSF in the case of aberration-free and out-of-focus positioning in 1885 [21,22]. In 1942, Nijboer expanded the aberration functions using Zernike polynomials, which allowed for the analytical evaluation of the PSF of small aberration optical systems [23]. Based on the research of Nijboer, Janssen et al. formulated a general expression using the power-Bessel series, solving the PSF for optical systems with large aberrations [24,25]. In 2019, Miks derived analytical formulas that enable the approximate calculation of the PSF for optical systems affected by defocus and spherical aberration [26].

This paper presents the first derivation of an analytical solution for the defocused PSF in diffraction-limited systems. The solution is expressed as a confluent hypergeometric function with rapid converging series expression. The findings exhibit strong consistency across various defocusing scenarios. Further exploration of this analytical solution can yield insights into the image degradation of diverse design configurations within diffraction-limited systems under defocusing conditions, thereby establishing a unified model for high-definition imaging systems in proximity to the focal point.

2. Fresnel Diffraction and Fourier–Bessel Transform of Converging Spherical Waves

The fundamental function of an incoherent optical imaging system is to transform the spherical wave that emerges from a point source on the object plane through the entrance pupil so that it converges into a focused wave and is directed onto the image plane via the exit pupil. In the case of a diffraction-limited system without any geometric aberrations, these converging waves are spherical. However, if any geometric aberration exists, the wavefront of the converging wave deviates from being perfectly spherical; we will only discuss those cases where such aberrations are absent.

Assuming that a spherical wave undergoes diffraction through a circular exit pupil (coordinated in the ξ-η plane), it converges at point P $(0, 0, z_0)$, as illustrated in Figure 1.

When observed in the x-y coordinate system on the observation plane located at a distance δ beyond point P, it results in defocusing. Here, δ represents the amount of defocus, typically where $\delta \ll z_0$.

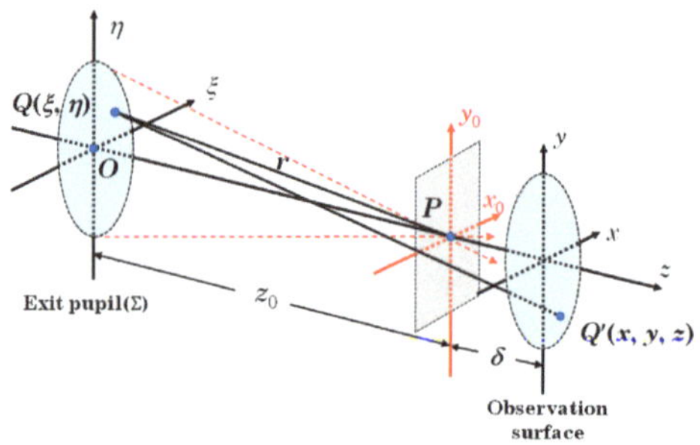

Figure 1. Schematic diagram of the defocus system.

The propagation effect of a light wave from the exit pupil to the observation plane is Fresnel diffraction. Considering a point denoted as Q (ξ, η) on the exit pupil and a point Q' (x, y, z) on the observation plane, according to the Fresnel diffraction theory, the field distribution ψ (x, y, z) is given by the following integral:

$$\psi(x,y,z) = \frac{e^{ikz}}{i\lambda z}\iint_{\Sigma} p(\xi,\eta)\exp\left\{i\frac{\pi}{\lambda z}\left[(x-\xi)^2+(y-\eta)^2\right]\right\}d\xi d\eta, \quad (1)$$

where p (ξ, η) is the pupil function, k is $2\pi/\lambda$, and λ is the wavelength.

Working from Figure 1, let r be the distance between point Q and point P on the ideal image plane axis. This yields the following approximate expression:

$$r = \left[z_0^2 + \left(\xi^2+\eta^2\right)\right]^{1/2} = z_0\left[1+\frac{\xi^2+\eta^2}{z_0^2}\right]^{1/2} \approx z_0 + \frac{\xi^2+\eta^2}{2z_0}. \quad (2)$$

The pupil function is the complex amplitude distribution of converging spherical waves on the pupil surface (see Figure 2), given by:

$$p(\xi,\eta) = \begin{cases} \frac{e^{-ikr}}{r} \approx \frac{1}{z_0}\exp\left[-i\frac{\pi}{\lambda z_0}\left(\xi^2+\eta^2\right)\right] & \sqrt{\xi^2+\eta^2} < a, \\ 0 & \sqrt{\xi^2+\eta^2} \geq a, \end{cases} \quad (3)$$

where a is the radius of the exit pupil.

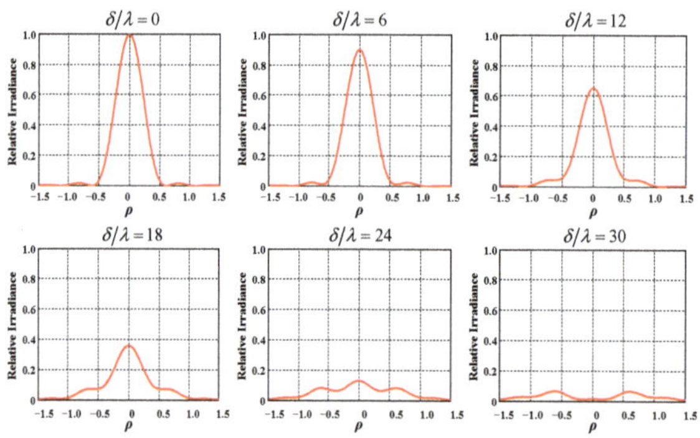

Figure 2. The PSF distribution under different degrees of defocusing.

Substituting the complex amplitude distribution $p(\xi,\eta)$ of the light wave on the exit pupil plane into the Fresnel diffraction formula, as shown in Equation (1), leads to complex amplitude distribution on the observation plane (defocused image plane) x-y:

$$\begin{aligned}\psi(x,y,z) = &\frac{e^{ikz}}{i\lambda z z_0}\exp\left[i\frac{\pi}{\lambda}\left(\frac{x^2+y^2}{z}\right)\right]\times \\ &\iint_{\Sigma}\exp\left[-i\frac{\pi\delta}{\lambda z_0(z_0+\delta)}(\xi^2+\eta^2)\right]\exp\left[-i\frac{2\pi}{\lambda z}(x\xi+y\eta)\right]d\xi d\eta.\end{aligned} \tag{4}$$

According to the rotational symmetry of the system, polar coordinates are used and normalized:

$$\begin{aligned}&\xi = ar\cos\theta, \eta = ar\sin\theta, \frac{x}{\lambda z} = \frac{\rho}{a}\cos\phi, \frac{y}{\lambda z} = \frac{\rho}{a}\sin\phi, \\ &C = \frac{e^{ikz}}{i\lambda z z_0}\exp\left[i\frac{\pi}{\lambda z}(x^2+y^2)\right], \kappa = \pi A^2(\delta/\lambda), A = a/z_0,\end{aligned} \tag{5}$$

where A is the numerical aperture, and r and ρ are the normalized dimensionless radius in the spatial domain (exit pupil coordinates) and frequency domain (observation plane coordinates), respectively. Thus, Equation (4) can now be expressed as:

$$\begin{aligned}\psi(x,y,z) &= \psi(\rho) \\ &= C\int_0^{2\pi}\left[\int_0^1 \exp\left(i\frac{\kappa}{(1+\delta/z_0)}r^2\right)rdr\right]\exp[-i2\pi r\rho\cos(\theta-\phi)]d\theta \\ &\propto \int_0^{2\pi}\left[\int_0^1 \exp(i\kappa r^2)rdr\right]\exp(-i2\pi r\rho\cos\theta)d\theta \\ &= \mathfrak{B}\left\{\int_0^1 \exp(i\kappa r^2)rdr\right\}.\end{aligned} \tag{6}$$

Because the $\delta/z_0 \ll 1$, δ/z_0 is omitted in the first-order approximation, and so is the constant, C. Equation (6) indicates that due to the rotation symmetry of the system, the complex amplitude distribution $\psi(\rho)$ on the defocusing plane is a Fourier–Bessel transform, denoted by symbol B of the integral $\int_0^1 \exp(i\pi\kappa r^2)rdr$, with the parameter κ under the first-order approximation.

The integral expression of the zero-order Bessel function is:

$$J_0(z) = \frac{1}{2\pi}\int_0^{2\pi} e^{iz\cos\theta}d\theta, \tag{7}$$

Substituting it into Equation (6) by setting $z = 2\pi r\rho$, one obtains:

$$\begin{aligned}\psi(A,\delta;\rho) &= 2\pi\int_0^1 \exp\left[i\kappa r^2\right]J_0(2\pi r\rho)rdr \\ &= 2\pi\int_0^1 \exp\left[i\pi A^2(\delta/\lambda)r^2\right]J_0(2\pi r\rho)rdr.\end{aligned} \tag{8}$$

3. Analytical Expression of the PSF

3.1. Zero-Order Approximate (Focal Plane) PSF—Airy Disk

When δ = 0 (with no defocus), the intensity distribution (PSF) corresponding to Equation (8) is the Airy disk:

$$h(\rho) = |\psi(\rho)|^2 = \left[2\pi\int_0^1 J_0(2\pi r\rho)rdr\right]^2 = \left[\frac{J_1(2\pi\rho)}{\rho}\right]^2, \tag{9}$$

Let $\mu_{1,1}$ be the first zero point of J_1 (x), then:

$$\rho = \frac{\mu_{1,1}}{2\pi} = \frac{3.832}{2\pi} = 0.61. \tag{10}$$

It can be established that the radius of the first dark ring of the Airy disk is:

$$x = \rho\frac{\lambda z_0}{a} = 0.61\frac{\lambda}{A}. \tag{11}$$

3.2. First-Order Approximation (Defocus) PSF—Confluent Hypergeometric Function

When $\delta \neq 0$, one can obtain from Equation (8):

$$\begin{aligned} h(A,\delta;\rho) &= \left|2\pi a^2 C \int_0^1 \exp(i\kappa r^2) J_0(2\pi r\rho) r dr\right|^2 \\ &\propto (2\pi)^2 \left\{ \left[\int_0^1 \cos(\kappa r^2) J_0(2\pi r\rho) r dr\right]^2 + \left[\int_0^1 \sin(\kappa r^2) J_0(2\pi r\rho) r dr\right]^2 \right\}. \end{aligned} \tag{12}$$

The series expression of the Bessel function of order zero is as follows:

$$J_0(x) = \sum_{n=0}^{\infty} \frac{(-1)^n}{(n!)^2} \left(\frac{x}{2}\right)^{2n}. \tag{13}$$

Substituting Equation (13) into Equation (12), setting $u = r^2$ and omitting the unimportant constants before the integral, yields:

$$\begin{aligned} h(A,\delta;\rho) &= \left[\sum_{n=0}^{\infty} \frac{(-1)^n (\pi\rho)^{2n}}{(n!)^2} \left[\int_0^1 \cos(\kappa u) u^n du\right]\right]^2 + \left[\sum_{n=0}^{\infty} \frac{(-1)^n (\pi\rho)^{2n}}{(n!)^2} \left[\int_0^1 \sin(\kappa u) u^n du\right]\right]^2 \\ &= \left[\sum_{n=0}^{\infty} \frac{(-1)^n (\pi\rho)^{2n}}{(n!)^2} I_C^{\langle n\rangle}\right]^2 + \left[\sum_{n=0}^{\infty} \frac{(-1)^n (\pi\rho)^{2n}}{(n!)^2} I_S^{\langle n\rangle}\right]^2, \end{aligned} \tag{14}$$

where $I_C^{\langle n\rangle}$, $I_S^{\langle n\rangle}$ can be written as:

$$\begin{cases} I_C^{\langle n\rangle} = \int_0^1 \cos(\kappa u) u^n du = \frac{1}{2(n+1)}[F(n+1,n+2;i\kappa) + F(n+1,n+2;-i\kappa)], \\ I_S^{\langle n\rangle} = \int_0^1 \sin(\kappa u) u^n du = \frac{-i}{2(n+1)}[F(n+1,n+2;i\kappa) - F(n+1,n+2;-i\kappa)], \end{cases} \tag{15}$$

and where $F(\alpha,\gamma;z)$ is the confluent hyper-geometric function [27,28], defined as follows:

$$\begin{aligned} F(\alpha,\gamma;z) &= \sum_{k=0}^{\infty} \frac{(\alpha)_k}{k!(\gamma)_k} z^k \\ &= 1 + \frac{\alpha}{\gamma}\frac{z}{1!} + \frac{\alpha(\alpha+1)}{\gamma(\gamma+1)}\frac{z^2}{2!} + \frac{\alpha(\alpha+1)(\alpha+2)}{\gamma(\gamma+1)(\gamma+2)}\frac{z^3}{3!} + \cdots. \end{aligned} \tag{16}$$

Thus, the analytical solution of the defocus PSF of the diffraction-limited imaging system can be presented as:

$$\begin{aligned} h(A,\delta;\rho) = &\left\{\sum_{n=0}^{\infty} \frac{(-1)^n (\pi\rho)^{2n}}{[2(n+1)^2 n!]^2} [F(n+1,n+2;i\kappa) + F(n+1,n+2;-i\kappa)]\right\}^2 + \\ &\left\{\sum_{n=0}^{\infty} \frac{(-1)^n (\pi\rho)^{2n}}{[2(n+1)^2 n!]^2} [F(n+1,n+2;i\kappa) - F(n+1,n+2;-i\kappa)]\right\}^2. \end{aligned} \tag{17}$$

For $\alpha = n+1$ and $\gamma = n+2$, these functions can be expressed as series that converge very rapidly:

$$\begin{cases} F(n+1,n+2;i\kappa) + F(n+1,n+2;-i\kappa) = 2(n+1)\sum_{k=0}^{\infty} \frac{(-1)^k}{(n+2k+1)} \frac{\kappa^{2k}}{(2k)!}, \\ F(n+1,n+2;i\kappa) - F(n+1,n+2;-i\kappa) = -2i(n+1)\sum_{k=1}^{\infty} \frac{(-1)^k}{(n+2k)} \frac{\kappa^{2k-1}}{(2k-1)!}. \end{cases} \tag{18}$$

Substituting Equation (18) into Equation (17) yields a series expression for the PSF:

$$h(A,\delta;\rho) = \left[\sum_{n=0}^{\infty} (-1)^n \frac{(\pi\rho)^{2n}}{(n!)^2} \sum_{k=0}^{\infty} \frac{(-1)^k \kappa^{2k}}{(n+2k+1)(2k)!}\right]^2 + \left[\sum_{n=0}^{\infty} (-1)^n \frac{(\pi\rho)^{2n}}{(n!)^2} \sum_{k=1}^{\infty} \frac{(-1)^k \kappa^{2k-1}}{(n+2k)(2k-1)!}\right]^2. \tag{19}$$

When $A = 0.242$ and $\delta/\lambda = 0,6,12,18,24,30$, the PSF is shown in Figure 2. It can be seen that by increasing the defocus, the brightness of the central point of the PSF decreases, and more energy diffuses to the sidelobes, resulting in increased spreading.

In evaluating the convergence speed of the series equation, Equation (19), for a system with typical parameter values, seven sampling values on the PSF curve are taken into account, as depicted in Figure 3. The abscissa represents the truncation term number N. All terms from $n = 0$ to $n = \mathrm{N}$ are retained and are included in the summation, while terms with $n > \mathrm{N}$ are omitted. The ordinate represents the corresponding values of the PSF curves with labels. The various curves in the graph represent the convergence status of the functions at the corresponding positions in the small graph. Evidently, all the curves converge rapidly and approach constants when N > 12, indicating that the confluent hypergeometric function series converges smoothly and quickly.

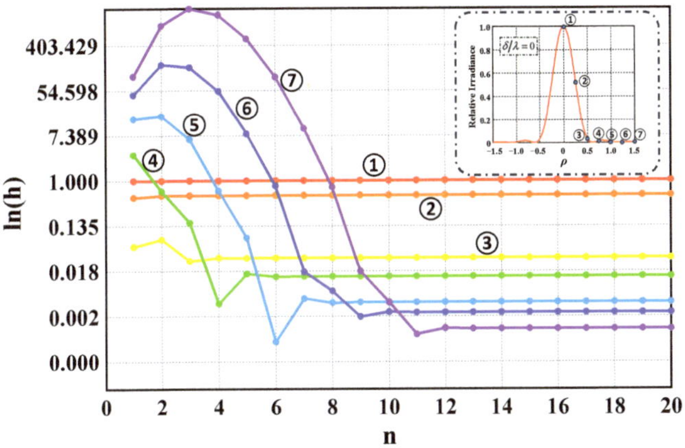

Figure 3. The sampling values of the PSF curve vs. the term's truncation number N of the series equation, Equation (19). All terms from $n = 0$ to $n = \mathrm{N}$ are kept, but the terms with $n > \mathrm{N}$ are omitted.

3.3. *Optical Transfer Functions for Diffraction-Limited Systems with Defocus*

The optical transfer function (OTF) is a crucial, complex-valued function that describes the response of an imaging system in relation to spatial frequency; it is a crucial parameter in analyzing and evaluating the performance of imaging systems. The OTF of the diffraction limit system is expressed as:

$$\begin{aligned} H(\nu) = \; & \tfrac{1}{\pi\kappa\nu}\cos\left(4\kappa\nu^2\right)\{J_1(4\kappa\nu)\cos^{-1}(\nu) - \\ & \sum_{n=0}^{\infty}\tfrac{(-1)^n}{2n}\left[J_{2n-1}(4\kappa\nu) - J_{2n+1}(4\kappa\nu)\right]\sin\left[2n\cos^{-1}(\nu)\right] + \\ & \tfrac{1}{2\kappa\nu}\sum_{n=0}^{\infty}(-1)^n 2nJ_{2n}(4\kappa\nu)\sin\left[2n\cos^{-1}(\nu)\right]\} + \\ & \tfrac{1}{\pi\kappa\nu}\sin\left(4\kappa\nu^2\right)\left\{\sum_{n=0}^{\infty}\tfrac{(-1)^n}{2n-1}\left[J_{2n-2}(4\kappa\nu) - J_{2n}(4\kappa\nu)\right]\sin\left[(2n-1)\cos^{-1}(\nu)\right] - \right. \\ & \left. \tfrac{1}{2\kappa\nu}\sum_{n=0}^{\infty}(-1)^n(2n-1)J_{2n-1}(4\kappa\nu)\sin\left[(2n-1)\cos^{-1}(\nu)\right]\right\}, \end{aligned} \tag{20}$$

where $J_n(z)$ is the Bessel function of the first kind of the order n. Each subfigure in Figure 2 corresponds to an OTF curve in Figure 4. The OTF gradually decreases at each frequency as the degree of defocus increases.

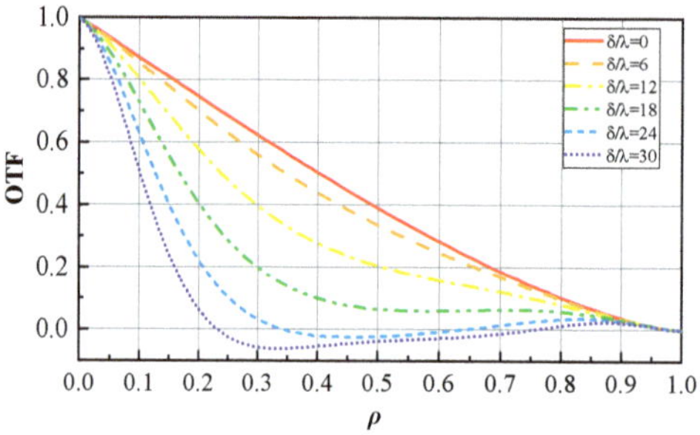

Figure 4. OTF curves for different defocus states.

4. The Characteristic Parameters of the Solution

4.1. Equivalent Bandwidth and Equivalent Linewidth

Both the OTF and the PSF are complex functions, for which it is convenient to introduce a single parameter to characterize the image quality for comparison. Goodman proposed equivalent bandwidth and equivalent linewidth as evaluation indexes [17]. We can assume that the main components of the signal (the image and its spatial spectrum) in both the spatial and frequency domains are concentrated only in a finite region near the center. To measure the spreading of the function $g(x)$ and its bandwidth $G(\nu)$, the relation between the function $g(x)$ and $G(\nu)$ is given by Fourier transform:

$$\begin{cases} G(\nu) = \int_{-\infty}^{+\infty} g(x)\exp(-2i\pi\nu x)dx, \\ g(x) = \int_{-\infty}^{+\infty} G(\nu)\exp(2i\pi\nu x)d\nu. \end{cases} \tag{21}$$

It can be established from the above two equations that:

$$G(0) = \int_{-\infty}^{+\infty} g(x)dx,\ g(0) = \int_{-\infty}^{+\infty} G(\nu)d\nu. \tag{22}$$

Equation (22) has a clear geometric meaning: if the height of a rectangle is equal to $G(0)$ and the area is the same as the area under the curve $G(\nu)$, its width is called "equivalent bandwidth"; see Figure 5a. The reciprocal of the slope of the object image's light intensity curve at the object's edge is called the "equivalent line width"; see Figure 5b. The equivalent bandwidth and equivalent linewidth can be expressed as follows:

$$\tilde{\omega} = \frac{\int_{-\infty}^{\infty} G(\nu)d\nu}{G(0)},\ \tilde{s} = \frac{\int_{-\infty}^{\infty} g(x)dx}{g(0)}. \tag{23}$$

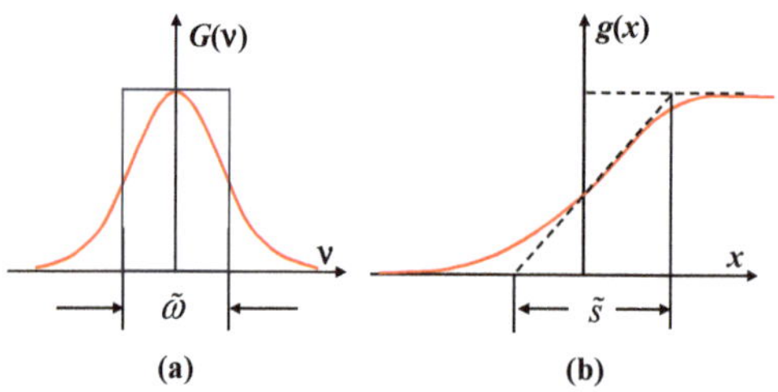

Figure 5. The characteristic parameter (**a**) Equivalent bandwidth and (**b**) equivalent linewidth.

The product of equivalent linewidth and equivalent bandwidth can be obtained from Equations (22) and (23):

$$\tilde{s}\tilde{\omega} \approx 1. \tag{24}$$

The above equation represents the mutual restriction between the signal spreading in the spatial domain and that in the frequency domain, which is also called "uncertainty".

Due to the symmetry, both equations $H(\alpha, v)$ and $h(\alpha, \rho)$ have rotationally symmetrical surfaces. One may make a cylinder where the top surface is at a tangent to the surface and where the bottom surface coincides with the coordinate plane, such that the volume of the cylinder is equal to the volume under the surface (Figure 6). It is physically reasonable to take the diameter $2R$ of the cylinder as either the equivalent bandwidth of the OTF or the equivalent linewidth of the PSF.

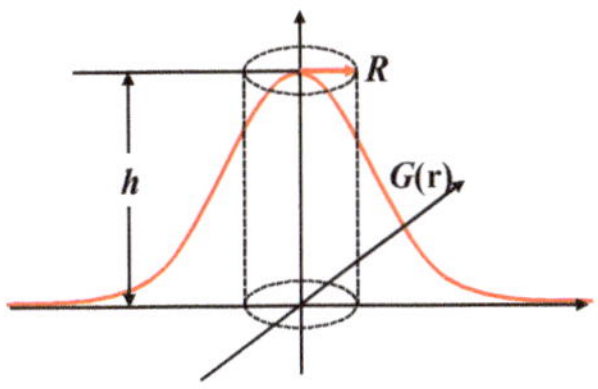

Figure 6. Schematic of the two-dimensional structure of the equivalent bandwidth.

When the surface expression is $G(r)$, then the volume of the cylinder is:

$$\pi R^2 h = \int_0^{2\pi} d\varphi \int_0^{\infty} G(r) r dr = 2\pi \int_0^{\infty} G(r) r dr. \tag{25}$$

From Equation (25), the following result is obtained:

$$R = \sqrt{2\int_0^{\infty} G(r) r dr / h} = \sqrt{\int_{-\infty}^{\infty} G(r) r dr / h}, \tag{26}$$

where h is the maximum value of the surface; that is the center value.

When defocus occurs, the equivalent bandwidth will become smaller, while the equivalent linewidth will become larger, indicating increased spreading. By substituting the expression for the OTF (Equation (20)) and the expression for the PSF (Equation (19)) into Equation (26), a series of equivalent bandwidth and equivalent linewidth values are obtained, as depicted in Table 1.

Table 1. Equivalent linewidth $\tilde{s}$ and equivalent bandwidth $\tilde{\omega}$ values at defocus.

δ/λ	$\tilde{s}$ (μm)	$\tilde{\omega}$ (mm^{-1})	$\tilde{s}\tilde{\omega}$
0	2.27	440	0.998
6	2.39	418	0.999
12	2.80	356	0.997
18	3.74	265	0.991

The last row of the above table shows that, corresponding to a given amount of defocus, the product of equivalent linewidth and equivalent bandwidth is approximately equal to 1.

4.2. Strehl Ratio and Resolution

The Strehl ratio (SR) is a measure of the relative peak intensity of an optical system's PSF, compared to that of an ideal, aberration-free system. In Equation (19), the SR can be calculated by setting $n = 0$, which requires considering only the term $\rho = 0$ in the following equation:

$$SR = h(A, \delta; \rho)|_{\rho=0} = \left[\sum_{k=0}^{\infty} \frac{(-1)^k \kappa^{2k}}{(2k+1)!}\right]^2 + \left[\sum_{k=1}^{\infty} \frac{(-1)^k \kappa^{2k-1}}{(2k)!}\right]^2. \tag{27}$$

This formula can be simplified to a trigonometric function:

$$SR(A,\delta) = \left[\frac{2\sin(\kappa/2)}{\kappa}\right]^2. \quad (28)$$

In Figure 7, the red curve represents the confluent hypergeometric function, and the blue is calculated by Equation (28); their values are highly consistent. The values of each point are shown in Table 2.

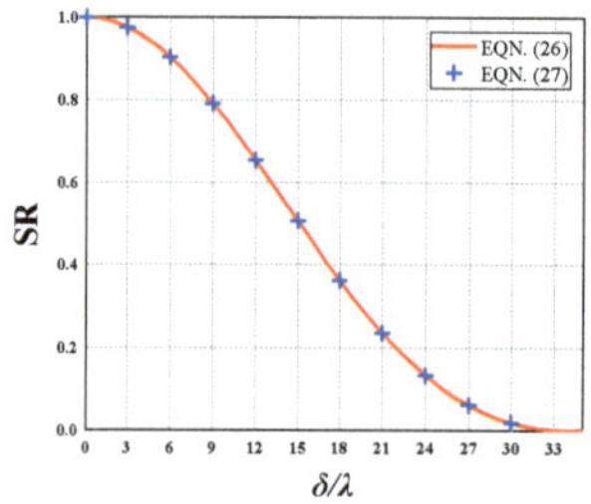

Figure 7. Diagram of the SR as a function of the defocus amount.

Table 2. Defocus amount versus the SR.

δ/λ	0	3	6	9	12	15	18	21	24	27	30
SR	1	0.975	0.902	0.791	0.654	0.506	0.362	0.235	0.133	0.061	0.018

The relationship between the SR and resolution will be discussed in the following content. Resolution typically denotes the maximum spatial frequency that can be distinguished. Within industry settings, the resolution of the design output when the OTF falls between 0.15 and 0.2 is commonly employed as a reference value.

Figure 8 shows a series of defocus OTF curves, obtained with an optical system numerical aperture of 0.242 and a main wavelength of 550 nm, the defocus range being 0~21 times the wavelength. The intersection points between these curves and the line where OTF = 0.2 are regarded as the resolution in each defocus state.

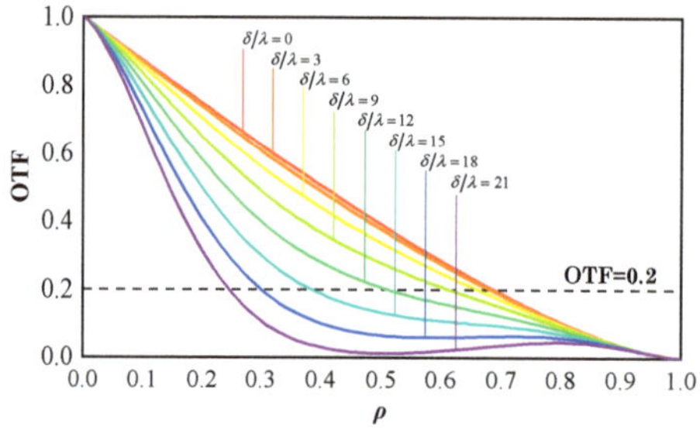

Figure 8. OTF versus frequency.

Figure 9a shows the curve of spatial resolution as a function of the SR. The abscissa of each discrete point represents the intersection point between the curve of different defocus states (see Figure 8) and the line OTF = 0.2, which is the resolution. The ordinate is the SR value at the corresponding defocus position. The ordinate in Figure 9b is the logarithm of the SR; the results show that the resolution is more closely linearly related to ln(SR).

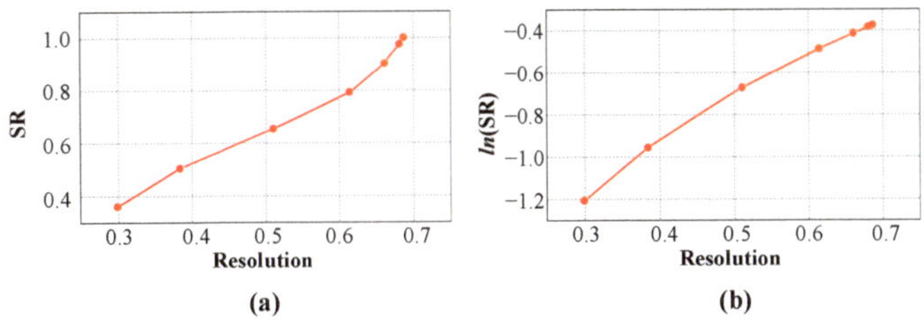

Figure 9. The curve of the relationship between the SR and resolution at OTF = 0.2 for various defocus positions. (**a**) SR vs. resolution; (**b**) ln(SR) vs. resolution.

5. Comparison of the Analytical Solution with the Ray-Tracing Results of the Real System

To verify the accuracy of our analytical results, we implemented ray tracing with the optical design software, Zemax (version 18.4.1), to generate both the FFT PSF and FFT MTF. The FFT used here refers to the fast Fourier transform technique, which is the method employed by Zemax software for calculating the PSF and MTF of optical systems. The MTF is the absolute value of the OTF, known as the modulus of the OTF. The calculation method for the FFT PSF is as follows. A set of grid rays emitted from the source point propagate to the exit pupil; the complex amplitude of the corresponding point on the wavefront grid at the exit pupil is calculated using the amplitude and path difference. After appropriate scaling and Fourier transformation, the square of the resultant values yields the real value of the PSF. The FFT MTF essentially represents the Fourier transform of the FFT PSF. In subsequent content, these will be referred to simply as PSF and MTF.

The 2D system layout of the three systems used for comparison is illustrated in Figure 10. Light rays from the left side of the system converge on the image plane on the right side.

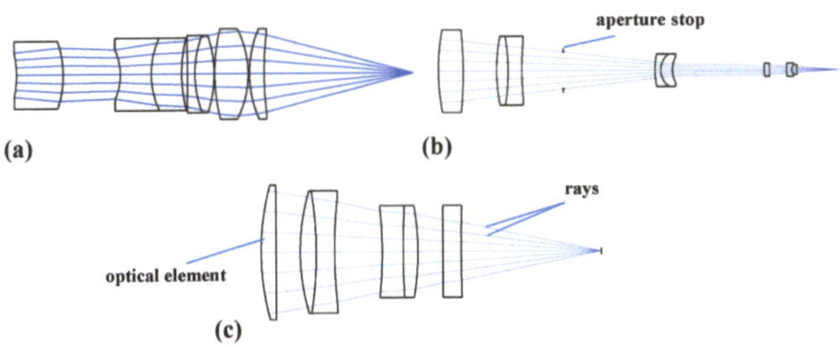

Figure 10. The 2D system layouts: (**a**) system 1; (**b**) system 2; (**c**) system 3.

Table 3 lists the main optical parameters of systems 1–3. For systems 1–3, which have respective A values of 0.2425, 0.0698, and 0.1961, step values of δ/λ were chosen as 5, 50, and 5 to accurately reflect the variations in MTF and PSF with defocus.

Table 3. Main parameters of the optical systems.

System	Wavelength	$F/\#$	NA'	f'	Diameter (Entrance Pupil)	RMS/Waves
1	588 nm	2.00	0.2425	20	10.0	0.0039
2	550 nm	7.14	0.0698	500	70.0	0.0124
3	550 nm	2.54	0.1961	100	40.0	0.0005

We compared the results obtained by the Zemax software's ray tracing with the PSF and MTF calculated by the analytical solution. The defocused MTF and PSF curves of the three systems are shown in Figures 11–13, respectively, where the solid line represents the result calculated by the analytical solution and the dashed line represents the result obtained by ray tracing. Table 4 presents the root mean square error (RMSE) values calculated between the MTF and PSF, obtained through two different methods at different defocus levels.

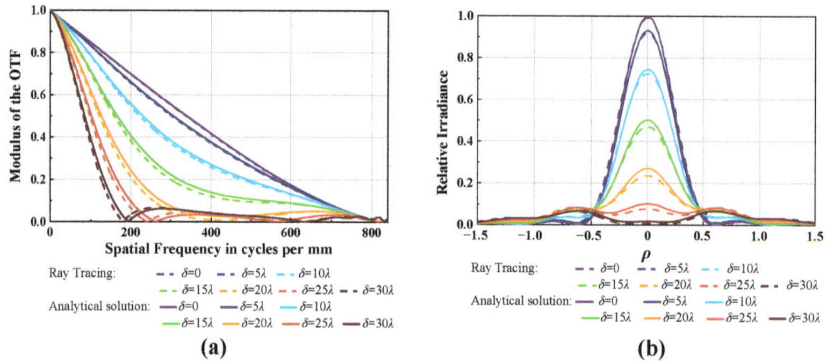

Figure 11. Comparison of the analytical solution and ray tracing results for system 1 at different defocus levels. (**a**) MTF; (**b**) PSF.

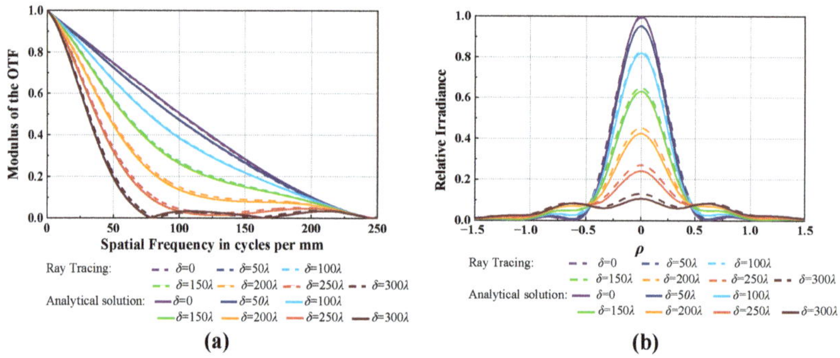

Figure 12. Comparison of the analytical solution and ray tracing results for system 2 at different defocus levels. (**a**) MTF; (**b**) PSF.

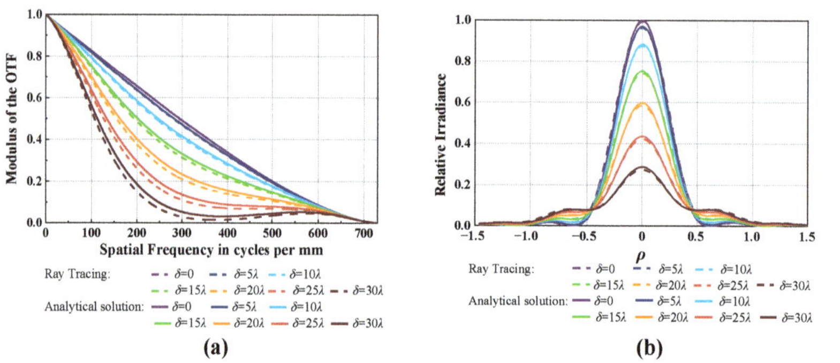

Figure 13. Comparison of the analytical solution and ray tracing results for system 3 at different defocus levels. (**a**) MTF; (**b**) PSF.

Table 4. Comparison of the PSF and the MTF for different systems.

	System 1 (RMSE)		System 2 (RMSE)		System 3 (RMSE)	
MTF	$\delta = 0$	0.25%	$\delta = 0$	0.33%	$\delta = 0$	0.08%
	$\delta = 5\lambda$	0.39%	$\delta = 50\lambda$	0.23%	$\delta = 5\lambda$	0.22%
	$\delta = 10\lambda$	0.92%	$\delta = 100\lambda$	0.27%	$\delta = 10\lambda$	0.60%
	$\delta = 15\lambda$	1.48%	$\delta = 150\lambda$	0.63%	$\delta = 15\lambda$	1.10%
	$\delta = 20\lambda$	1.72%	$\delta = 200\lambda$	0.94%	$\delta = 20\lambda$	1.57%
	$\delta = 25\lambda$	1.54%	$\delta = 250\lambda$	1.05%	$\delta = 25\lambda$	1.89%
	$\delta = 30\lambda$	1.34%	$\delta = 300\lambda$	0.97%	$\delta = 30\lambda$	2.01%
PSF	$\delta = 0$	1.22%	$\delta = 0$	1.23%	$\delta = 0$	1.23%
	$\delta = 5\lambda$	1.04%	$\delta = 50\lambda$	1.22%	$\delta = 5\lambda$	1.16%
	$\delta = 10\lambda$	0.76%	$\delta = 100\lambda$	1.24%	$\delta = 10\lambda$	1.00%
	$\delta = 15\lambda$	0.95%	$\delta = 150\lambda$	1.31%	$\delta = 15\lambda$	0.79%
	$\delta = 20\lambda$	1.17%	$\delta = 200\lambda$	1.35%	$\delta = 20\lambda$	0.60%
	$\delta = 25\lambda$	0.99%	$\delta = 250\lambda$	1.25%	$\delta = 25\lambda$	0.50%
	$\delta = 30\lambda$	0.56%	$\delta = 300\lambda$	0.97%	$\delta = 30\lambda$	0.48%

6. Discussion

Many imaging systems are precision-designed and fabricated to achieve imaging quality that is close to the diffraction limit, in order to deliver exceptional performance. These systems include cameras equipped in aerospace vehicles, high-end plan-apochromatic microscope objectives used in medicine and life sciences, object identification systems, and imaging systems employed in semiconductor applications. For high-end imaging systems, the most commonly encountered imaging error is defocusing rather than inherent aberrations. Defocusing may be caused by variations in object distance, fabrication and assembly tolerances, environmental temperature fluctuations, and other factors.

By applying the above confluent hypergeometric functions, this paper offers an analytic solution for the Fresnel diffraction of a diffraction-limited system. The universal model describes the behaviors of systems under defocusing. The analytic model's most compelling feature is its complete independence from intricate design details, objective types, and various application scenarios. It is anticipated that such an analytic formula will facilitate the analysis of image quality deterioration, the establishment of suitable tolerances for design and fabrication, and the prediction of defocusing effects.

7. Conclusions

In this paper, the analytic formula of the defocused PSF of the diffraction-limited system utilizing confluent hypergeometric function is first derived, and then the characteristic parameters, including equivalent bandwidth, equivalent linewidth, and the Strehl ratio, are derived. The analytical expression of the Strehl ratio of defocusing systems can be reduced to a simple trigonometric function. Examples of three different types of diffraction-limited systems are analyzed and compared with the analytic solution. The PSF of each example is very close to the analytic solution, and the S values of defocus are consistent with those calculated from the above analytical expressions.

It follows that the defocused PSF of the diffraction-limited system, which includes the confluent hypergeometric function, is indeed a typical expression of a diffraction-limited system. This analytical formula can give the defocus characteristics and behavior of all diffraction-limited systems with known or unknown structural parameters. This analytical solution describes the properties of all diffraction-limit systems, including defocus performance, resolution, etc.

Author Contributions: Conceptualization, F.S., Q.C., X.T. and F.X.; methodology, F.S., Q.C. and X.T.; software, F.S., Q.C. and X.T.; validation, F.S. and Q.C.; formal analysis, F.S. and Q.C.; investigation, F.S., Q.C., X.T. and F.X.; data curation, F.S. and Q.C.; writing—original draft preparation, F.S., Q.C. and X.T.; writing—review and editing, F.S., Q.C., X.T. and F.X.; visualization, Q.C.; project administration,

X.T. and F.X.; funding acquisition, X.T. and F.X. All authors have read and agreed to the published version of the manuscript.

Funding: This research was funded by the National Key Research and Development Program of China (2021YFB3601400).

Institutional Review Board Statement: Not applicable.

Informed Consent Statement: Not applicable.

Data Availability Statement: Data are contained within the article.

Conflicts of Interest: The authors declare no conflicts of interest.

References

1. Born, M.; Wolf, E. Elements of the theory of diffraction. In *Principles of Optics*, 7th ed.; Cambridge University Press: Cambridge, UK, 2019; Volume 8, pp. 412–516.
2. Song, F.; Jutamulia, S. Fourier analysis and image quality evaluation of imaging systems. In *Modern Optical Information Processing*, 2nd ed.; Peking University Press: Beijing, China, 2014; Volume 6, pp. 108–141.
3. Meitav, N.; Ribak, E.N.; Shoham, S. Point Spread Function Estimation from Projected Speckle Illumination. *Light Sci. Appl.* **2015**, *5*, e16048. [CrossRef]
4. Jin, X.; Liu, L.; Chen, Y.; Dai, Q. Point Spread Function and Depth-Invariant Focal Sweep Point Spread Function for Plenoptic Camera 20. *Opt. Express* **2017**, *25*, 9947. [CrossRef] [PubMed]
5. Long, M.; Soubo, Y.; Weiping, N.; Feng, X.; Jun, Y. Point-Spread Function Estimation for Adaptive Optics Imaging of Astronomical Extended Objects. *Astrophys. J.* **2020**, *888*, 20. [CrossRef]
6. Moradi, S.; Moallem, P.; Sabahi, M.F. Scale-Space Point Spread Function Based Framework to Boost Infrared Target Detection Algorithms. *Infrared Phys. Technol.* **2016**, *77*, 27–34. [CrossRef]
7. Chen, S.; Feng, H.; Pan, D.; Xu, Z.; Li, Q.; Chen, Y. Optical Aberrations Correction in Postprocessing Using Imaging Simulation. *ACM Trans. Graph.* **2021**, *40*, 1–15. [CrossRef]
8. Chen, S.; Lin, T.; Feng, H.; Xu, Z.; Li, Q.; Chen, Y. Computational Optics for Mobile Terminals in Mass Production. *IEEE Trans. Pattern Anal. Mach. Intell.* **2023**, *45*, 4245–4259. [CrossRef] [PubMed]
9. Qiao, C.; Li, D.; Liu, Y.; Zhang, S.; Liu, K.; Liu, C.; Guo, Y.; Jiang, T.; Fang, C.; Li, N.; et al. Rationalized Deep Learning Super-Resolution Microscopy for Sustained Live Imaging of Rapid Subcellular Processes. *Nat. Biotechnol.* **2023**, *41*, 367–377. [CrossRef] [PubMed]
10. Chan, D.; Sheinin, M.; O'Toole, M. SpinCam: High-Speed Imaging via a Rotating Point-Spread Function. In Proceedings of the 2023 IEEE/CVF International Conference on Computer Vision (ICCV), Paris, France, 2–3 October 2023; IEEE: Paris, France, 2023; pp. 10755–10765. [CrossRef]
11. Shechtman, Y.; Sahl, S.J.; Backer, A.S.; Moerner, W.E. Optimal Point Spread Function Design for 3D Imaging. *Phys. Rev. Lett.* **2014**, *113*, 133902. [CrossRef] [PubMed]
12. Li, H.; Wang, F.; Wei, T.; Miao, X.; Cheng, Y.; Pang, X.; Jiang, K.; Huang, W.; Zhang, Y. Particles 3D Tracking with Large Axial Depth by Using the 2π-DH-PSF. *Opt. Lett.* **2021**, *46*, 5088. [CrossRef] [PubMed]
13. Mahmoud, A.; Xu, D.; Xu, L. Optical Design of High Resolution and Shared Aperture Electro-Optical/Infrared Sensor for UAV Remote Sensing Applications. In Proceedings of the 2016 IEEE International Geoscience and Remote Sensing Symposium (IGARSS), Beijing, China, 10–15 July 2016; IEEE: Beijing, China, 2016; pp. 2921–2924. [CrossRef]
14. Chang, Y.-S.; Hsu, L.; Huang, K.-L. Adjustable-Focus Ultracompact Endoscopic Lens Design with Ultrahigh Optical Performance. *Appl. Opt.* **2018**, *57*, 1091. [CrossRef] [PubMed]
15. Kirshner, H.; Aguet, F.; Sage, D.; Unser, M. 3-D PSF Fitting for Fluorescence Microscopy: Implementation and Localization Application. *J. Microsc.* **2013**, *249*, 13–25. [CrossRef] [PubMed]
16. Diaz Zamboni, J.; Casco, V. Estimation Methods of the Point Spread Function Axial Position: A Comparative Computational Study. *J. Imaging* **2017**, *3*, 7. [CrossRef]
17. Goodman, J.W. *Introduction to Fourier Optics*, 3rd ed.; Roberts & Company Publishes: Greenwood, IN, USA, 2005.
18. Stavroudis, O.N. *The Optics of Rays, Wavefronts, and Caustics*; Academic Press: New York, NY, USA, 1972.
19. Stavroudis, O.N. *The Mathematics of Geometrical and Physical Optics: The K-Function and Its Ramifications*, 1st ed.; Wiley: Hoboken, NJ, USA, 2006. [CrossRef]
20. Cain, S.C.; Watts, T. Nonparaxial Fourier Propagation Tool for Aberration Analysis and Point Spread Function Calculation. *Opt. Eng.* **2016**, *55*, 085104. [CrossRef]
21. Lommel, E. Diffraction by a slot and strips. *Abh. Bayer. Akad.* **1885**, *15*, 233–328.
22. Lommel, E. Theory and experimental investigations of diffraction phenomena at a circular aperture and obstacle. *Abh. Bayer. Akad.* **1886**, *15*, 531.
23. Nijboer, B.R.A. The Diffraction Theory of Aberrations. Ph.D. Thesis, University of Groningen, Groningen, The Netherlands, 1942.

24. Braat, J.; Dirksen, P.; Janssen, A.J.E.M. Assessment of an Extended Nijboer–Zernike Approach for the Computation of Optical Point-Spread Functions. *J. Opt. Soc. Am. A* **2002**, *19*, 858. [CrossRef]
25. Janssen, A.J.E.M. Extended Nijboer–Zernike Approach for the Computation of Optical Point-Spread Functions. *J. Opt. Soc. Am. A* **2002**, *19*, 849. [CrossRef]
26. Mikš, A.; Novák, J. Point Spread Function of an Optical System with Defocus and Spherical Aberration—Analytical Formulas. *Appl. Opt.* **2019**, *58*, 5823. [CrossRef] [PubMed]
27. Andrews, L.C. *Special Functions of Mathematics for Engineers*, 2nd ed.; SPIE Press: Bellingham, WA, USA, 1998.
28. Gradshteyn, I.S.; Ryzhik, I.M. *Table of Integrals, Series, and Products*, 5th ed.; Academic Press: Cambridge, UK, 1980.

photonics

MDPI

Article

Pulse Duration Compression by Two-Stage Stimulated Brillouin Scattering and Stimulated Raman Scattering

Xiaoyan Han [1], Zhaohong Liu [1,2,*], Shaowen Li [1], Sensen Li [1,3], Zhongwei Yang [1], Qiang Su [1], Yu Zhang [2], Bayanheshig [4], Yuanqin Xia [1], Yulei Wang [1] and Zhiwei Lu [1]

1 Hebei Key Laboratory of Advanced Laser Technology and Equipment, Tianjin 300401, China; 202021901004@stu.hebut.edu.cn (X.H.); 173757@stu.hebut.edu.cn (S.L.); sensli@163.com (S.L.); 202221901035@stu.hebut.edu.cn (Z.Y.); 2018134@hebut.edu.cn (Q.S.); xiayq@hebut.edu.cn (Y.X.); wyl@hebut.edu.cn (Y.W.); zhiweilv@hebut.edu.cn (Z.L.)

2 Shanxi Key Laboratory of Advanced Semiconductor Optoelectronic Devices and Integrated System, Jincheng 048026, China; zhangyu@semi.ac.cn

3 Science and Technology Electro-Optical Information Security Control Laboratory, Tianjin 300308, China

4 Changchun Institute of Optics, Fine Mechanics and Physics, Chinese Academy of Sciences, Changchun 130033, China; bayin888@sina.com

* Correspondence: lzh@hebut.edu.cn

Abstract: A pulse duration compression technique that combines stimulated Brillouin scattering (SBS) and stimulated Raman scattering (SRS) was presented in this study, achieving an output pulse duration of 48.3 ps. The feasibility of this approach has been experimentally demonstrated. To be specific, a pulse duration of 7.4 ns is compressed to 48.3 ps with an energy of 5.27 mJ, and the energy efficiency of the SRS pulse duration compression system is up to 21.84%. Moreover, this study provides a practical method for reliably generating high-energy short pulses.

Keywords: stimulated Brillouin scattering; stimulated Raman scattering; pulse duration compression; high-energy short pulse

1. Introduction

High-energy short-pulse laser generation has been a current research hotspot due to its unique characteristics of short pulse duration and high peak power, which have good application prospects in fields including laser processing [1,2], laser ranging [3,4] and laser ignition [5,6]. At present, there are several methods for obtaining short-pulse lasers, such as Q-switched technology [7,8] and mode-locked technology [9–11]. Q-switched technology is used for generating high-energy short pulses with low cost, but it can only output sub-nanosecond pulses due to the size limitations of the gain medium imposed by the cavity length. Mode-locked technology enables the output of picosecond or even femtosecond pulses of several μJ. Further amplification suffers from deterioration of the beam quality and damage to the optical element [12].

In addition to the above-mentioned schemes, utilizing optical nonlinear effects to achieve pulse peak-power enhancement and energy amplification has become an effective way to overcome the limitations in the power enhancement process. One of the effects is that the pulse compression technology based on stimulated Brillouin scattering (SBS) exhibits significant advantages in achieving short pulses owing to its phase conjugation, high energy efficiency and simple structure [13–15]. It has been demonstrated to be an effective technology for generating short pulses with energy up to the joule level [16]. Hon et al. proposed and practically realized the pulse duration compression of SBS in gas and used CH_4 gas to compress the pump pulse duration from 20 ns to 3 ns in the tapered waveguide structure [17]. Liu et al. employed a focused single-cell structure to compress the pump pulse from 8.05 ns and 145 mJ to 1.35 ns and 94.5 mJ, and the energy efficiency reached 65% [18]. SBS pulse compression technology can compress nanosecond laser pulses to sub-nanosecond levels. Further compression for achieving shorter pulse duration output has become a primary area of research focus. However, when the laser wavelength is 1064 nm, the application of SBS pulse duration compression technology is constrained by the theoretical compression limit of phonon lifetime, making it difficult to obtain short pulses within 100 ps.

The SRS pulse compression technology provides a wide gain bandwidth, making it more possible to acquire picosecond pulses, and it has been reported to successfully compress pulses on the picosecond scale [19,20]. Chevalier et al. achieved a high backward SRS efficiency of 40% by selecting acetone as the medium under the condition of the maximum pump energy, which was 1 mJ [21]. Despite its small compression limit, the wider application of SRS is hindered by its low energy efficiency. Currently, enhancing the energy efficiency of SRS pulse compression remains a significant challenge. SBS pulse compression technology is characterized by its high energy efficiency, but it is unable to achieve the generation of short pulses within the 100 ps range due to the limit of the theoretical compression imposed by the phonon lifetime. Additionally, SRS pulse compression technology has a smaller compression limit, while its energy efficiency is lower. SBS pulse duration compression can efficiently compress nanosecond pulses to hundreds of picoseconds and provide a high-power pump for SRS compression. In this case, a narrow pulse duration can lower the gain of nonlinear effects such as SBS and improve the efficiency of SRS. It is an effective method to obtain short pulses by combining them for pulse compression, aiming to overcome their respective shortcomings.

In this study, an effective pulse compression method is proposed based on the combination of SBS and SRS. The sub-nanosecond short pulses obtained through the SBS are the input of the SRS. The short pulse within 100 ps is output, and the energy efficiency of SRS is improved. The experiment finally outputs short pulses with a pulse duration of 48.3 ps, and the energy efficiency of the SRS reaches 21.84%.

2. Theoretical Analysis

During the amplification process, the backward Stokes generated in the SRS generator is taken as the seed to extract pump energy. The low energy efficiency of SRS is caused by the existence of a high-order Stokes, which is excited by a first-order Stokes after reaching a certain energy level. Only the first-order Stokes absorbs energy for amplification, while the high-order Stokes does not participate in the amplification process. To improve energy efficiency, it provides a feasible way to raise the first-order generation threshold, which makes the high-order Stokes less likely to occur. The generation threshold formula is shown below [22]:

$$\mathrm{P_{th}} = \frac{\mathrm{T} + 2\alpha\mathrm{L}}{2\mathrm{gL}}\pi\omega_{\mathrm{p}}^2(0) \tag{1}$$

where T is the transmission of the output coupler; α refers to the absorption coefficient of the medium; L denotes the length of the medium; g signifies the Raman gain coefficient of the medium; and $\omega_{\mathrm{p}}(0)$ means the waist radius of the pump beam at the center of the medium. The expression for the gain coefficient of SRS is given by Equation (2) [23]:

$$g = \frac{\omega_{-1}\chi^{(3)}T_2\gamma_v}{4n^2\mathrm{A}c^2\omega_v} \tag{2}$$

where ω_{-1} is the angular frequency of the light wave; $\chi^{(3)}$ denotes the third-order nonlinear polarizability; T_2 means the phonon lifetime; γ_v refers to the dispersion response of nonlinear polarization intensity; n indicates the refractive index; A signifies the amplitude of the electric field; c presents the speed of light in a vacuum; and ω_v is the angular frequency of phonon vibration. When $X = \omega_0 \chi^{(3)} T_2 \gamma_v / 4n^2 c^2 \omega_v$, according to the light intensity expression $I_i = 0.5n E_0 c A^2$, Equation (2) can be expressed as follows:

$$g = X\sqrt{nc\varepsilon_0/2I_i} \tag{3}$$

where ε_0 is the permittivity of free space. Based on the relationship $I = E/\tau s$ between light intensity and incident energy, Equation (3) can be expressed as:

$$g = X\sqrt{nc\varepsilon_0 \tau s/2E} \tag{4}$$

Substituting Equation (4) into Equation (1) yields:

$$P_{th} = \frac{(T + 2\alpha L)\pi\omega_p^2(0)\sqrt{E}}{LX\sqrt{2nc\delta_0\tau s}} \tag{5}$$

In Figure 1, the relationship between the input pulse duration and the generation threshold determined through numerical simulation is illustrated. In the SRS pulse compression, the backward Stokes generated in the generator meets the pump light in the amplifier, and the Stokes generated in the generator is used as the seed light to extract the pump energy because the higher-order Stokes are stimulated by the accumulation of first-order Stokes energy to a certain extent. In addition, the existence of the higher-order Stokes inevitably consumes the first-order Stokes. The higher-order Stokes does not participate in the amplification process, and only the first-order Stokes extracts the pump energy to be compressed and amplified, causing a lower energy efficiency. To improve the energy efficiency of SRS pulse compression, a feasible method involves raising the threshold of first-order Stokes, and thus its energy accumulation cannot reach levels that would excite higher-order Stokes. As can be seen from the red dashed circle in Figure 1, a substantial change occurs in the threshold when the input pulse duration is 0.6 ns or less. Therefore, to enhance the energy efficiency of the SRS, the input pulse duration should be shorter than 0.6 ns. With this adjustment, the intensity of the generated Stokes experiences a significant improvement, effectively fulfilling the objective of enhancing energy efficiency.

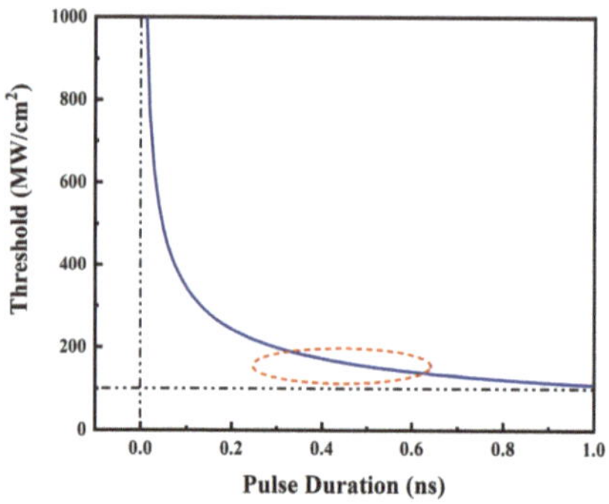

Figure 1. The relationship between threshold and input pulse duration.

3. Experiment Setup

The experimental setup diagram of the SBS and SRS combined compression system is depicted in Figure 2. The pump is a passively Q-switched Nd: YAG laser [24], and its output is the Gaussian waveform with a wavelength of 1064 nm, a pulse duration of 7.4 ns and a beam quality factor of $M^2 = 1.2$. P_1, P_2, FR and HP_1 form a spatial optical isolator. HP_2 and P_3 are combined to form an energy regulator that controls the pumping of energy into the SBS compression. In the current experiment, the pump pulse output is detected by the laser, and a time-domain waveform with a pulse duration of 7.4 ns is obtained. The details are illustrated in Figure 2. The medium used in the SBS compression is FC-40 (g = 1.6 cm/GW, τ = 0.3 ns), and a compact double-cell structure is employed to output 28.6 mJ of Stokes with a pulse duration of 580 ps at a pump energy of 35.2 mJ. On this basis, SRS compression is conducted, and the SRS adopts a generation–amplification double-cell structure with a medium of $Ba(NO_3)_2$ crystal whose length is 5 cm (g = 11 cm/GW@1064 nm, Δv = 0.4 cm^{-1}). After being compressed by SBS, a portion of the output light is reflected by the beam splitter (BS) and focused by the lens (F_2) on the generator cell to generate backward Stokes, serving as seed light during the amplification process. However, the other portion is transmitted as the pump through the BS and then meets the Stokes seed in the amplifier through the half-wave plate (HP_4) and polarizer (P_5). The seed light extracts energy from the pump for compression and amplification, and the Stokes is ultimately output through the dichroic mirror (SP_2). Additionally, energy is recorded using an energy meter (OPHIR PE50BB-DIF-C and OPHIR NOVA II); pulse duration characteristics are measured using an ultrafast phototube (UPD-35-IR2-D, Alphalas; rise time < 35 ps, wavelength range: 800–1700 nm) combined with a digital oscilloscope (DPO71254C, Tektronix; bandwidth: 12.5 GHz, sampling rate: 100 Gs/s); and a beam quality analyzer (CCD, Dataray, Redding, CA, USA) is used to detect and explore the spatial profile of the beam.

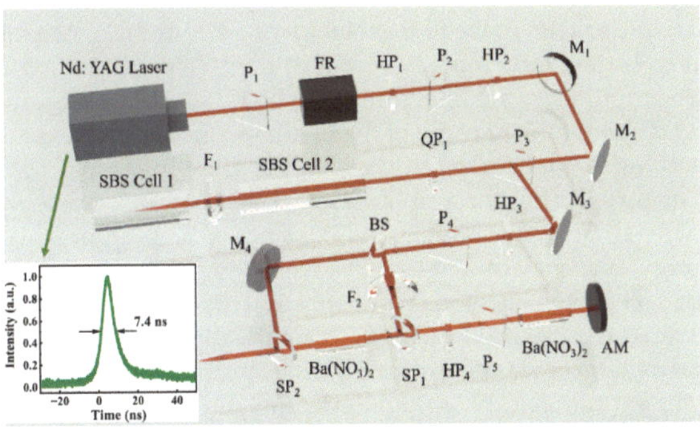

Figure 2. Schematic diagram of the experimental setup of the SBS and SRS combined compression system. The illustration represents the pump waveform output by the laser. P: polarizer; FR: Faraday rotator; HP: half-wave plate; M: high-reflective mirror; QP: quarter-wave plate; F: lens; BS: beam splitter; SP: dichroic mirror (HR@1064 nm, HT@1198 nm); AM: absorb mirror.

4. Results and Discussion

The output characteristics of SBS compression are measured. As shown in Figure 3a, the output energy and energy efficiency are recorded simultaneously. With the increase in input energy, the output energy increases linearly and the corresponding energy efficiency initially rises rapidly at low input energy levels and later reaches a stable value. At the low energy range of 5–30 mJ, with the increase in input energy, the energy efficiency exhibits a steep upward trend, and the pump energy extracted by seed light is compressed and amplified by SBS. When it is characterized by higher energy, the extraction speed decreases and the energy efficiency slowly increases. Then, the pumping energy extracted by Stokes seed light reaches a saturation point. Finally, the energy efficiency stabilizes at approximately 75%. The evolution of the output pulse duration in relation to the input energy is shown in Figure 3b. With the increase in input energy, the output pulse duration first decreases to the minimum value and subsequently increases slowly. The reason for this phenomenon is that the Stokes leading edge absorbs pump energy to saturation, and then, with the continuous increase in input energy, the leading edge is in a saturated state and no longer absorbs energy, while the trailing edge absorbs energy and is amplified, resulting in a phenomenon in which the output Stokes pulse duration is widened. At a pump energy of 35.2 mJ, the output Stokes pulse duration is 580 ps and the corresponding output energy is 26.8 mJ.

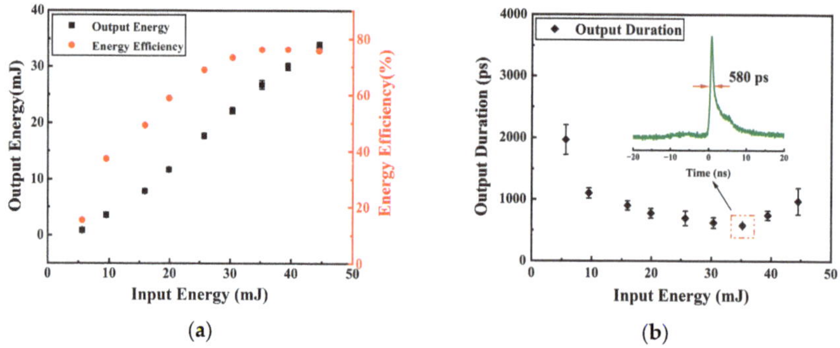

Figure 3. (**a**) Measured output energy of pulses compressed via SBS (black squares) and the corresponding energy efficiency (red circles) with respect to input energy; (**b**) output duration evolution of the Stokes radiation with respect to the input energy (inset: a typical compressed waveform); black rhombuses measure SBS duration.

On this basis, the input characteristics of the SRS generator are determined. A stable output of Stokes is observed when the input energy reaches 1.5 mJ. The generation of SRS is cascaded, where higher-order Stokes are directly excited by first-order Stokes after reaching a certain energy level. Therefore, the energy of the first-order Stokes is consumed subsequently by the generated high-order Stokes. Moreover, the first-order Stokes acts as the seed to extract pump energy during the amplification process of SRS, while the higher-order Stokes is not involved, further leading to the existence of high-order Stokes significantly affecting the energy efficiency of first-order Stokes. To improve the energy efficiency of the SRS generator, an effective method is to ensure that the high-order Stokes is not generated in the generator. The use of short pulses with a pulse duration of 580 ps as the input for SRS generator has a high generation threshold, which prevents the first-order Stokes energy from reaching the excitation level for high-order Stokes.

As displayed in Figure 4, the spectrum of the Stokes output by the generator under various input energies was recorded, including values such as 1.5 mJ, 4.87 mJ and 5.39 mJ. Only the first-order Stokes is present in Figure 4a,b, while in Figure 4c, the second-order Stokes appears at an energy of 5.39 mJ. The maximum input energy used in the experiment for the generator is 4.87 mJ. Only the first-order Stokes with a wavelength of 1198 nm is present under this condition.

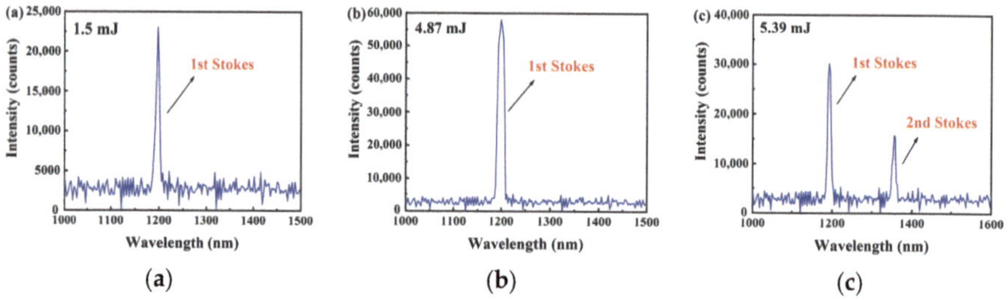

Figure 4. The spectrum diagrams of output Stokes from the generator with different input energies: (**a**) 1.5 mJ; (**b**) 4.87 mJ; (**c**) 5.39 mJ.

In the experiment, the output pulse duration is detected by an ultrafast phototube connected to a digital oscilloscope with a bandwidth of 12.5 GHz. As the theoretical detection limit of an oscilloscope is generally considered to be the reciprocal of the bandwidth, it can be concluded that the theoretical detection limit of the digital oscilloscope used in the experiment is 80 ps. Due to the theoretical detection limit, the rise time of the ultrafast photocell and the response time of the digital oscilloscope have a certain impact on the measured results. The actual pulse duration (τ_R) output by SRS pulse compression is calculated according to Equation (6) [25]:

$$\tau_R = \left(\tau_D^2 - \tau_P^2 - \tau_O^2\right)^{1/2} \quad (6)$$

where τ_D is the result of the experiment; τ_P refers to the rising time of the ultrafast photoelectric probe; and τ_O denotes the response time of the digital oscilloscope. Based on the experimental measurement results, the actual pulse duration output by the generator and the amplifier is calculated according to the above formula.

As shown in Figure 5a, the output energy of the generator has been rapidly increasing with the rise in input energy, and the corresponding energy efficiency has also been increasing. With the increase in input energy, the energy of backward Stokes generated in the generator also increases. The output energy is 0.56 mJ and the energy efficiency is 11.6%, while the input energy is 4.87 mJ. As the actual output pulse duration is affected by the bandwidth of the oscilloscope and the photoprobe, the actual output pulse duration is obtained by subtracting the bandwidth of both from the pulse duration displayed by the oscilloscope. The measured and actual results of the generator are respectively shown in Figure 5b,c. There is no significant fluctuation in the output pulse duration within the range of the input energy of the generator cell. As can be seen in Figure 5b,c, the pulse waveforms of insets are classified into positive and negative signals. In general, it is shown that the positive signal values are the normal measured optical signals, and the negative signal values are the electromagnetic oscillation caused by a short falling edge of the pulse waveform. Therefore, the area with positive signal values is considered the pulse duration. When the input energy is 4.87 mJ, the actual output pulse duration is 64.3 ps.

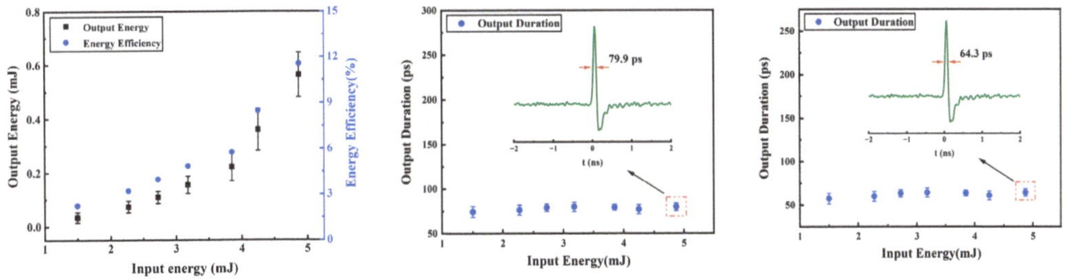

(a) (b) (c)

Figure 5. Output characteristics of the SRS generator: (**a**) output energy (black squares) and energy efficiency (blue circles) of Stokes pulses versus input pump energy; (**b**) output duration evolution of Stokes with respect to input energy (the measured result); the inset is the pulse waveform with an input energy of 4.87 mJ; (**c**) output duration evolution of Stokes with respect to input energy (the actual result); the inset is the pulse waveform with an input energy of 4.87 mJ.

Figure 6 shows the spatial distribution pattern of the light beam output by the SRS generator. It is measured at a position of 3 cm after leaving the generating cell. The transverse diameter and longitudinal diameter of the spatial beam profile are respectively marked as 3.13 and 2.65 mm in Figure 6a. The *z*-axis in Figure 6b only represents the intensity of light at varying positions. The color change on the right side of Figure 6b indicates that the light intensity, in turn, shifts from blue to red (from weak to strong), and the corresponding relative intensity changes from 0 to 1. The spatial profile of the beam output by the generator features a well-defined Gaussian distribution.

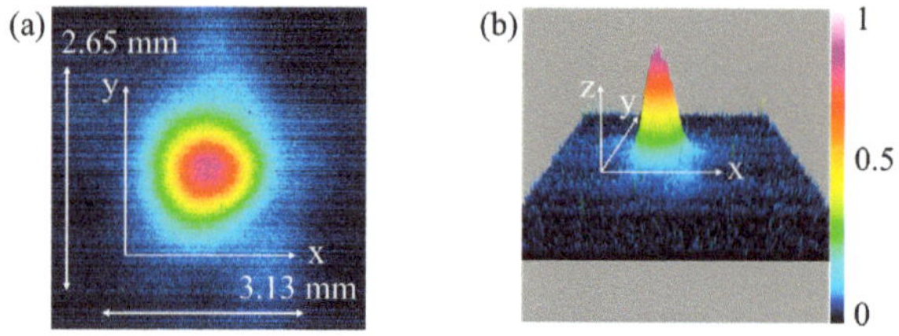

Figure 6. The spatial beam profile of the SRS generator: (**a**) plane graph; (**b**) schematic diagram of light intensity at different positions.

The input energy of the SRS generator is fixed at 4.87 mJ, with an output energy of 0.56 mJ, an energy efficiency of 11.6% and an output pulse duration of 64.3 ps. The Stokes pulse generated in this case serves as the seed in the amplifier. The 580 ps output by the SBS compression is taken as the pump in the SRS amplification process. The characteristics of the SRS amplifier are displayed in Figure 7. The output energy increases continuously with the increase in pump energy, as shown in Figure 7a. The energy efficiency first increases rapidly with the increase in pump energy, and, subsequently, it slows down while the extraction of pump energy by the Stokes seed gradually reaches saturation, ultimately stabilizing at around 24%. The evolution of the output pulse duration displayed by the oscilloscope is recorded in Figure 7b. With the increase in pump energy, the output pulse duration first decreases at a fast speed, and then its change is inconspicuous. When the pump energy is 2.79 mJ, the lower pump power limits the higher power seed from extracting energy. The pump power is already higher than the Stokes seed, while the pump energy reaches 4.35 mJ. In this case, the Stokes seed can effectively extract energy from the pump, leading to further compression of the output pulse. As the energy further increases, the extraction gradually reaches saturation and the output pulse duration tends to stabilize. The illustration in Figure 7b shows the time-domain waveform of the output pulse at an input energy of 19.26 mJ. In this case, the output pulse duration is 67.7 ps, consistent with an output energy of 4.71 mJ and an energy efficiency of 24.5%.

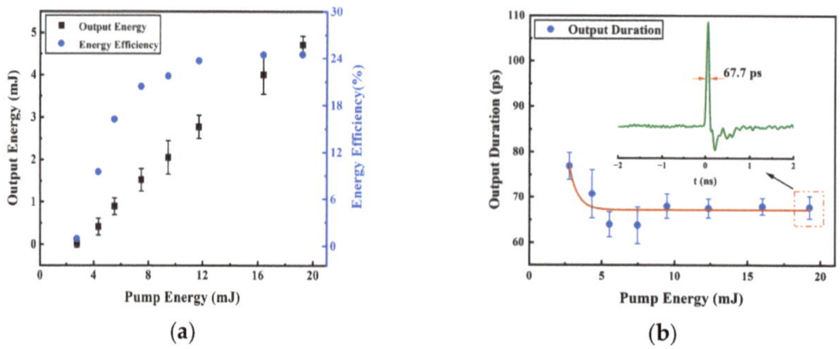

Figure 7. (**a**) Output energy (black squares) and energy efficiency (blue circles) of Stokes pulses versus pump energy of the SRS amplifier; (**b**) output pulse duration versus pump energy of the SRS amplifier; the inset is the pulse waveform with an energy of 19.26 mJ.

The characteristics of SRS compression are presented in Figure 8. Figure 8b illustrates the actual pulse duration, which is the pulse duration output by the oscilloscope minus the bandwidth influence of the oscilloscope and photoelectric probe. The horizontal axis indicates the total input energy of the combination of the SRS generator and the amplifier. A total output energy of 5.27 mJ, an energy efficiency of 21.84% and an actual output pulse duration of 48.3 ps are finally obtained at an input energy of 24.13 mJ.

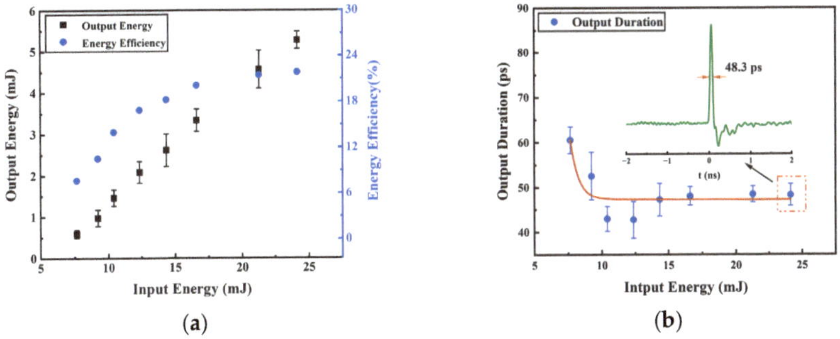

Figure 8. Output characteristics of the SRS compression: (**a**) output energy and energy efficiency of Stokes pulses versus input energy; (**b**) output pulse duration evolution of Stokes with respect to input energy.

The spatial profile of the final output Stokes from the SRS system is collected, as shown in Figure 9. It is measured at a position of 3 cm after leaving the generating cell. According to the observation, the transverse diameter and longitudinal diameter of the spatial profile section of the beam are respectively marked as 2.01 and 2.25 mm. The z-axis in Figure 9b only suggests the intensity of light at different positions. The color change on the right side of Figure 9b indicates that the light intensity, in turn, shifts from blue to red (from weak to strong), and the corresponding relative intensity varies from 0 to 1. Compared with the spatial beam profile output by the SRS generator, the distribution of radiant intensity output by the SRS generator–amplifier is more concentrated on the laser spot center. Therefore, during the SRS process, the center of the beam undergoes higher compression and amplification than the edge of the beam, thus increasing the concentration of energy distribution. The beam quality is measured, and a beam quality factor M^2 of 1.3 is obtained.

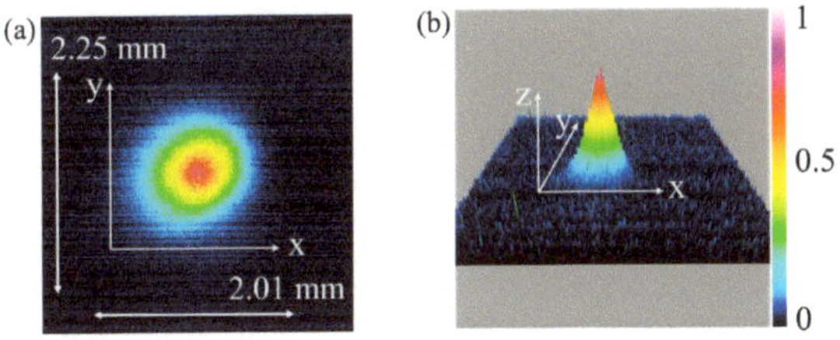

Figure 9. The spatial beam profile of the output Stokes from the SRS system: (**a**) plane graph; (**b**) schematic diagram of light intensity at different positions.

5. Conclusions

To conclude, SRS pulse compression achieved further compression of the sub-nanosecond short pulses output by SBS pulse compression. The technology of combining the SBS and SRS systems successfully obtained short-pulse outputs within 100 ps. The SRS system reached a total energy output of 5.27 mJ at a total input energy of 24.13 mJ, with an energy efficiency of up to 21.84% and a corresponding output pulse duration of 48.3 ps, which could successfully achieve short-pulse outputs within 100 ps at high energy levels and good beam quality. Through further optimization of experimental parameters, high-energy short pulses within the 100 ps range can be used in fields including precision instrument processing, laser ranging and laser spectroscopy.

Author Contributions: Conceptualization, Z.L. (Zhaohong Liu) and X.H.; Methodology, Z.L. (Zhaohong Liu) and X.H.; Software, Z.L. (Zhaohong Liu) and X.H.; Validation, S.L. (Shaowen Li) and Z.Y.; Formal Analysis, X.H.; Investigation, Z.L. (Zhaohong Liu); Resources, Z.L. (Zhaohong Liu); Data curation, Z.L. (Zhaohong Liu) and X.H.; Writing—original draft preparation, X.H.; Writing—review and editing, Z.L. (Zhaohong Liu) and X.H.; Visualization, X.H.; Supervision, Z.L. (Zhaohong Liu), S.L. (Sensen Li), Y.X., Y.W. and Z.L. (Zhiwei Lu); Project administration, Z.L. (Zhaohong Liu); Funding acquisition, Z.L. (Zhaohong Liu), S.L. (Sensen Li), Q.S., Y.Z., B., Y.X., Y.W. and Z.L. (Zhiwei Lu). All authors have read and agreed to the published version of the manuscript.

Funding: This research was funded by the National Natural Science Foundation of China (61905064, 61975050), the China Postdoctoral Science Foundation (300428), the Hebei Province Postdoctoral Special Grant (B2022005003), the Research Projects of High Education Institutions of Hebei Province (QN2019201), the Natural Science Research Foundation of Hebei University of Technology (JBKYXX2002), the Open Project Program of Shanxi Key Laboratory of Advanced Semiconductor Optoelectronic Devices and Integrated (2022SZKF04) and the National Key R&D Program of China (2022YFB3606100, 9813E2301X04).

Institutional Review Board Statement: Not applicable.

Informed Consent Statement: Not applicable.

Data Availability Statement: Data are contained within the article.

Conflicts of Interest: The authors declare no conflicts of interest.

References

1. Liu, H.; Sun, Z.; Chen, Y.; Zhang, W.; Chen, X.; Wong, C.-P. Laser Processing of Flexible In-Plane Micro-Supercapacitors: Progresses in Advanced Manufacturing of Nanostructured Electrodes. *ACS Nano* **2022**, *16*, 10088–10129. [CrossRef] [PubMed]
2. Jia, X.; Chen, Y.; Liu, L.; Wang, C.; Duan, J. Combined Pulse Laser: Reliable Tool for High-Quality, High-Efficiency Material Processing. *Opt. Laser Technol.* **2022**, *153*, 108209. [CrossRef]
3. Qian, G.; Xu, X.; Zhu, S.A.; Xu, C.; Gao, F.; Yakovlev, V.V.; Liu, X.; Zhu, S.Y.; Wang, D.W. Quantum Induced Coherence Light Detection and Ranging. *Phys. Rev. Lett.* **2023**, *131*, 033603. [CrossRef] [PubMed]
4. Lopac, N.; Jurdana, I.; Brnelić, A.; Krljan, T. Application of Laser Systems for Detection and Ranging in the Modern Road Transportation and Maritime Sector. *Sensors* **2022**, *22*, 5946. [CrossRef] [PubMed]
5. Pavel, N.; Bärwinkel, M.; Heinz, P.; Brüggemann, D.; Dearden, G.; Croitoru, G.; Grigore, O.V. Laser Ignition—Spark Plug Development and Application in Reciprocating Engines. *Prog. Quantum Electron.* **2018**, *58*, 1–32. [CrossRef]
6. Patane, P.; Nandgaonkar, M. Review: Multipoint Laser Ignition System and Its Applications to IC Engines. *Opt. Laser Technol.* **2020**, *130*, 106305. [CrossRef]
7. Chen, X.; Wang, N.; He, C.; Lin, X. Development of All-Fiber Nanosecond Oscillator Using Actively Q-Switched Technologies and Modulators. *Opt. Laser Technol.* **2023**, *157*, 108709. [CrossRef]
8. Jin, L.; Dai, W.; Yu, Y.; Dong, Y.; Jin, G. Single Longitudinal Mode Q-Switched Operation of Pr:YLF Laser with Pre-Lase and Fabry–Perot Etalon Technology. *Opt. Laser Technol.* **2020**, *129*, 106294. [CrossRef]
9. Dong, Z.; Lin, J.; Li, H.; Zhang, Y.; Gu, C.; Yao, P.; Xu, L. Er-Doped Mode-Locked Fiber Lasers Based on Nonlinear Polarization Rotation and Nonlinear Multimode Interference. *Opt. Laser Technol.* **2020**, *130*, 106337. [CrossRef]
10. Wu, J.W.; Liu, G.X.; Gao, Y.X.; Lin, X.B.; Cui, H.; Luo, Z.C.; Xu, W.C.; Likhachev, M.E.; Aleshkina, S.S.; Mashinsky, V.M.; et al. Switchable Femtosecond and Picosecond Spatiotemporal Mode-Locked Fiber Laser Based on NALM and Multimode Interference Filtering Effects. *Opt. Laser Technol.* **2022**, *155*, 108414. [CrossRef]
11. Wang, R.R.; Wang, Y.Y.; Dai, C.Q. Influence of Higher-Order Nonlinear Effects on Optical Solitons of the Complex Swift-Hohenberg Model in the Mode-Locked Fiber Laser. *Opt. Laser Technol.* **2022**, *152*, 108103. [CrossRef]
12. Zheng, S.; Chen, Z.; Lin, Q.; Cai, Y.; Lu, X.; Gao, Y.; Xu, S.; Fan, D. High-Gain Amplification for Femtosecond Optical Vortex with Mode-Control Regenerative Cavity. *Photon. Res. PRJ* **2020**, *8*, 1375–1380. [CrossRef]
13. Liu, Z.; Luo, T.; Chen, Y.; Li, S.; Li, S.; Fan, R.; Wang, Y.; Lu, Z. Generation of a High-Intensity Temporal Step Waveform Based on Stimulated Brillouin Scattering. *Photonics* **2022**, *9*, 309. [CrossRef]
14. Fan, R.; Liu, Z.; Jin, D.; Luo, T.; Li, N.; Li, S.; Wang, Y.; Xia, Y.; Lu, Z. High Temporal Waveform Fidelity Stimulated Brillouin Scattering Phase Conjugate Mirror Using Novec-7500. *Opt. Express OE* **2023**, *31*, 1878–1887. [CrossRef] [PubMed]
15. Wang, H.; Cha, S.; Kong, H.J.; Wang, Y.; Lu, Z. Sub-Nanosecond Stimulated Brillouin Scattering Pulse Compression Using HT270 for kHz Repetition Rate Operation. *Opt. Express OE* **2019**, *27*, 29789–29802. [CrossRef]
16. Feng, C.; Xu, X.; Diels, J.C. High-Energy Sub-Phonon Lifetime Pulse Compression by Stimulated Brillouin Scattering in Liquids. *Opt. Express OE* **2017**, *25*, 12421–12434. [CrossRef] [PubMed]
17. Hon, D.T. Pulse Compression by Stimulated Brillouin Scattering. *Opt. Lett. OL* **1980**, *5*, 516–518. [CrossRef]
18. Liu, H.; Li, Y.; Yang, C.; Gu, X.; Hu, W.; Zhang, Y.; Zhang, Y. 1.35 ns SBS Laser Pulse. *Optik* **2019**, *184*, 394–398. [CrossRef]
19. Kurbasov, S.V.; Losev, L.L. Raman Compression of Picosecond Microjoule Laser Pulses in $KGd(WO_4)_2$ Crystal. *Opt. Commun.* **1999**, *168*, 227–232. [CrossRef]

20. Warrier, A.M.; Lin, J.; Pask, H.M.; Mildren, R.P.; Coutts, D.W.; Spence, D.J. Highly Efficient Picosecond Diamond Raman Laser at 1240 and 1485 nm. *Opt. Express OE* **2014**, *22*, 3325–3333. [CrossRef]
21. Chevalier, R.; Sokolovskaia, A.; Tcherniega, N.; Rivoire, G. Stimulated Backward Raman Scattering Excited in the Picosecond Range: High Efficiency Conversions. *Opt. Commun.* **1991**, *82*, 117–122. [CrossRef]
22. Kitzler, O.; McKay, A.; Spence, D.J.; Mildren, R.P. Modelling and Optimization of Continuous-Wave External Cavity Raman Lasers. *Opt. Express OE* **2015**, *23*, 8590–8602. [CrossRef] [PubMed]
23. Makarov, N.S.; Bespalov, V.G. Effective Method of Anti-Stokes Generation by Quasi-Phase-Matched Stimulated Raman Scattering. *J. Opt. Soc. Am. B JOSAB* **2005**, *22*, 835–843. [CrossRef]
24. Cong, Z.; Liu, Z.; Qin, Z.; Zhang, X.; Wang, S.; Rao, H.; Fu, Q. RTP Q-Switched Single-Longitudinal-Mode Nd:YAG Laser with a Twisted-Mode Cavity. *Appl. Opt. AO* **2015**, *54*, 5143–5146. [CrossRef]
25. Wang, Y.; Gong, M.; Yan, P.; Huang, L.; Li, D. Stable Polarization Short Pulse Passively Q-Switched Monolithic Microchip Laser with [110] Cut Cr^{4+}:YAG. *Laser Phys. Lett.* **2009**, *6*, 788. [CrossRef]

hv **photonics**

MDPI

Communication

Combined Compression of Stimulated Brillouin Scattering and Laser–Induced Breakdown Enhanced with Sic Nanowire

Lai Feng [1,2], Yiming Zhao [2], Weiwei Zhang [2] and Dongsong Sun [1,*]

1 School of Earth and Space Sciences, University of Science and Technology of China, Hefei 230026, China; freyfl@126.com
2 Beijing Research Institute of Telemetry, Beijing 100076, China; zym_bird@126.com (Y.Z.); zhangww_511@163.com (W.Z.)
* Correspondence: sundongsong2023@163.com

Abstract: In this paper, the laser pulse time compression technique, based on stimulated Brillouin scattering (SBS) and passive laser–induced breakdown (LIB) series technology, is investigated. By doping a SiC nanowire in a CCl_4 solution of an LIB breakdown medium, the LIB generation threshold is reduced, and the stability of the LIB compression output is improved. When OD is 0.2, the output pulse width is 254.4 ps, and the corresponding energy conversion efficiency and pulse compression rate are 34.2% and 50.2%, respectively. Our experiment proves the feasibility of this scheme.

Keywords: SBS; LIB; the laser pulse time compression technique

1. Introduction

Ultrashort–pulse lasers are widely used in laser processing, spectral measurement, and high–energy physics [1–3]. The traditional method of generating an ultrashort–pulse laser through mode locking makes it difficult to generate a high–energy short–pulse laser due to the limitation of the damage threshold of the locking mirror [4–6]. Another method is to generate a nanosecond laser through Q–switching and then compress the laser to a picosecond level using SBS pulse–width compression technology, which has the advantages of a low cost, high energy conversion efficiency, phase conjugation, and high beam quality [7,8]. However, SBS pulse–width compression technology can only compress the pulse leading edge to obtain the laser pulse output in the order of a hundred picoseconds. In order to achieve a narrower pulse width, further post–pulse compression must be combined with other technical solutions.

The existing SBS and SRS series compression technology schemes [9,10], SBS and saturated gain switch series compression technology schemes [11], and SBS and LIB series compression technology schemes have been proven to be feasible [12]. The SBS and SRS series compression technology is the earliest technological combination to realize ultra–short pulses, which has been proved in a large number of studies, and that can achieve high repetition frequency output. However, after injecting the high peak power laser pulse output produced by the SBS pulse–width compression technology into the SRS medium, the SRS pulse–width compression technology will generate high–order Stokes light. The stability and energy conversion efficiency of the whole system are affected, making the scheme difficult to apply in high–power laser generation. The SBS and saturated gain switch series compression technology can generate a ps–level ultrashort–pulse laser, and the energy conversion efficiency is noteworthy. But, the saturated gain switch is only suitable for gas lasers, and this technology is only used in KrF lasers. The SBS and LIB series compression technology scheme combines pulse leading–edge compression technology with trailing–edge compression technology to yield complementary results. In addition, the high peak power Stokes light output from SBS pulse–width compression technology is injected into the LIB plasma switch, which is beneficial to the excitation of the plasma.

In 2017, Liu et al. [13] proposed a scheme of combining SBS and LIB compressions to simultaneously compress the leading and trailing edges of pulses, achieving a synchronous compression of the leading and trailing edges of the pulse and breaking through the limitation of SBS phonon lifetime. However, using pure water as a medium is affected by various nonlinear effects, while the LIB effect is not stable enough for practical applications. In this paper, CCl_4 is used as the medium to significantly improve the stability of LIB, and the stability and efficiency of the system are significantly enhanced by doping SiC nanowires. In addition, the commercial Faraday optical isolator is used to achieve optical isolation between the seed laser source and the amplifier. Meanwhile, the optical wedge is used to adjust the horizontal and vertical directions of the beam in the experimental setup, which improves the stability of the experimental setup and further improves the engineering application ability of the combined SBS and LIB pulse compression technology.

In this letter, in view of the limitation that SBS can only compress the leading edge of the injected light pulse, we proposed a scheme combining SBS leading–edge pulse compression technology with LIB trailing–edge compression technology in series and successfully realized the simultaneous compression of the leading and trailing edges of the injected light pulse. By doping the SiC nanowire into the CCl_4 solution to reduce the LIB threshold and improve the system stability, ultra–narrow pulse width and high–intensity laser outputs are finally achieved.

2. Material Preparation

The introduction of impure particles into a pure liquid medium can also significantly reduce the LIB threshold of the medium and improve the stability of the LIB [12]. LIB enhancement based on the thermal effect is a mechanism that enhances the performance of LIB through the doping of solid particles [14], where the selected doped solid particle materials are required to have a large difference in the thermal conductivity of the medium. The SiC material has significant nonlinear optical properties, excellent mechanical properties, and electrical conductivity. Its thermal conductivity is 100~200 W/m·K [15]. The SiC nanowire preparation method includes the following: Si and SiO_2 powders were mixed by ball milling according to the designed proportion (molar ratio of 1:1). The mixed Si and SiO_2 powders were placed at the bottom of the graphite crucible and covered with the porous carbon material. The graphite crucible was put into the center of the tube furnace, heated to 1400 °C at a rate of ~5 °C/min, and held at 1400 °C for 1 h under an argon gas environment.

The prepared SiC nanowire was characterized by electron microscopy, as shown in Figure 1. Figure 1a is a 1500× magnification image, and Figure 1b is a 4000× magnification image. SiC nanowire was integrated into the CCl_4 solution and dispersed homogeneously by ultrasound, and suspensions with OD values of 0.1, 0.15, and 0.2 were obtained, respectively. CCl_4 is a liquid organic solvent at room temperature. Its chemical properties are stable under natural conditions [16,17]. Then, suspensions with OD values of 0.1, 0.15, and 0.2 were injected into the LIB pool for narrow pulse–width compression experiments by combining SBS and LIB switches in series.

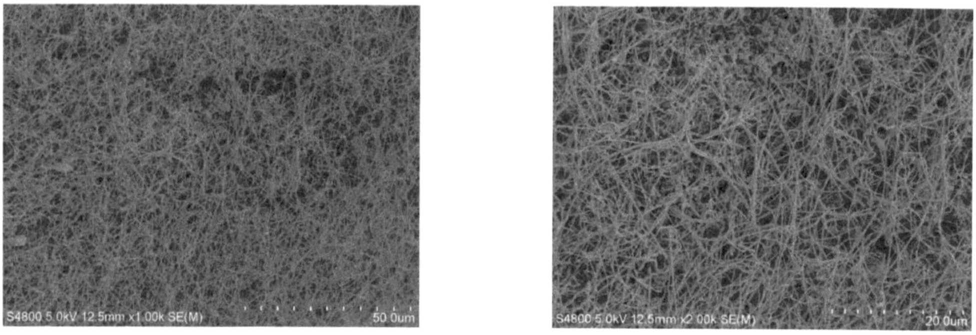

(**a**) (**b**)

Figure 1. SEM image of SiC nanowire: (**a**) 1500× magnification; (**b**) 4000× magnification.

3. Experiment

The experimental setup for the narrow–pulse width compression of a Nd:YAG single–longitudinal–mode laser using a structure of SBS and LIB switches in series is shown in Figure 2. During the experiment, the SBS is used for the first–stage compression; then, the LIB compression is performed with the steep rising–edge pulse output from the SBS. The Nd:YAG single–longitudinal–mode laser serves as a seed laser, with an output pulse width of 8 ns and a central wavelength of 1064 nm. The maximum output energy is 7 mJ at a repetition rate of 1 Hz. The optical Faraday isolator is used to avoid damage to the seed laser caused by backward light. The max beam diameter and isolation of the optical Faraday isolator are 9 mm and 40 dB, respectively. Lenses M_1 and M_2 are used to adjust the spot size of the optical Faraday isolator. Optical wedge pairs W_1~W_5 are used for beam adjustment, while right–angle prisms RM_1~RM_4 are used for beam reflection. Optical wedge pairs utilize the relative rotation of the two optical wedges. The output beam can be positioned in any direction in the corner cone with the input beam as the axis. Its advantages include an easy system assembly, stable structure, and easy realization of the engineering of optical systems. The deflection angle parameter of the optical wedge is 1°. The right–angle prism uses critical angle characteristics to achieve the efficient total reflection of the incident beam inside the right–angle prism. Compared with ordinary mirrors, the advantages of a right–angle prism include easy installation and high mechanical stress strength. A dual–pass amplifier structure is applied in the first–stage Nd:YAG amplifier. A half–wave plate and a polarizer P_1 are used to control the energy injected into the Nd:YAG amplifier crystal. Lenses M_3 and M_4 are used to adjust the beam diameter in the Nd:YAG amplifier crystal. A compact dual–pool compression structure is employed in the SBS compression section, including an amplification pool and a generation pool, both of which are 60 cm in length, and the SBS medium is FC–43. The FC–43 medium is one of the perfluorocarbon media. As an SBS medium, FC–43 has the characteristics of a high breakdown threshold, short phonon lifetime, large gain coefficient, and low absorption [18,19]. The phonon lifetime, SBS frequency shift, gain coefficient, and optical breakdown threshold of FC–43 medium are 200 ps, 1.3 cm/GW, 1073 MHz, and 178 GW/cm^2, respectively. In the single–pass amplification, the seed laser energy is amplified to 60 mJ. The amplified beam enters the amplification pool after passing through the quarter–wave waveplate and is focused into the generating pool by the lens M_5 with a focal length of 300 mm. The Stokes light generated near the focal area is compressed and amplified in the amplification pool by extracting the input energy in a backward direction. Stokes light is transformed into the *s*–polarized light through the quarter–wave plate and then further amplified in the Nd:YAG amplifier crystal, which is reflected by polarizer P_1 and the right–angle prism RM_3. The output beam passes through a half–wave plate and is incident to a type I phase–matched LiB_3O_5 (LBO) crystal for frequency doubling, producing green light in the 532 nm wavelength range. For the laser beam, the center wavelength, pulse width, and repetition rate are 1064 nm, 10 ns, and 10 Hz, respectively. The damage threshold of LBO crystal is better than 1 GW/cm^2. Dichroic mirror DM_1 is used to separate the 532 nm beam, and the 1064 nm beam is recovered using an optical trash can. Lens M_6 is used for focusing the 532 nm beam into the LIB generation pool with a length of 40 cm. The beam through the LIB pool is collimated by lens M_7. The forward–stimulated Raman component and compressed transmitted light are separated by a dichroic mirror DM_2 with high transmission for the 650 nm light and high reflectivity for the 532 nm light.

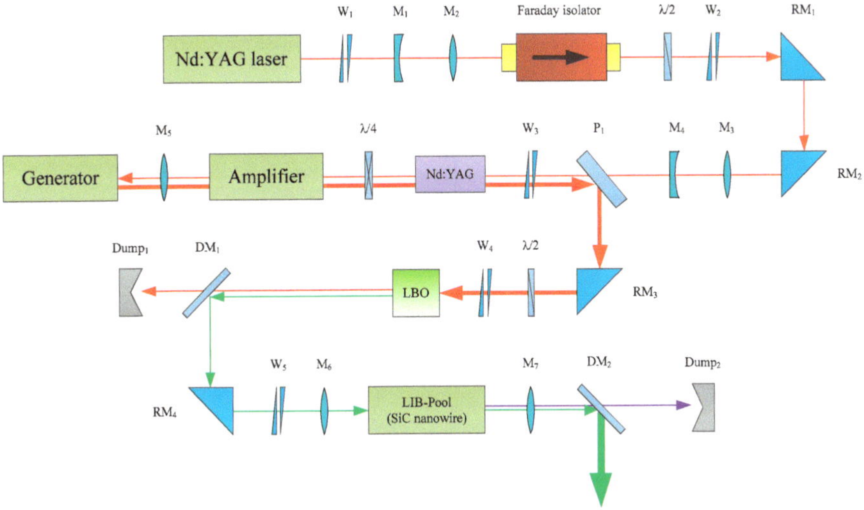

Figure 2. Optical path diagram of SBS and LIB switch series combination compression experiment.

4. Results and Discussion

The output Stokes beam compressed by SBS is focused into the LIB pool as a pump source for LIB compression after frequency doubling with LBO. After the occurrence of laser–induced plasma breakdown, the pulse width of the output waveform changes with the input energy, as shown in Figure 3. A total of 100 pulses recorded by a detection system using a photoelectric detector (Ultrafast UPD–50–UP) and a digital oscilloscope (Tektronix, DPO71254C, bandwidth: 12.5 GHz; sampling rate: 100 Gs/s) are captured for statistical analysis. The output laser energy is measured by an energy meter (ORHIR, PE50DIF–ER).

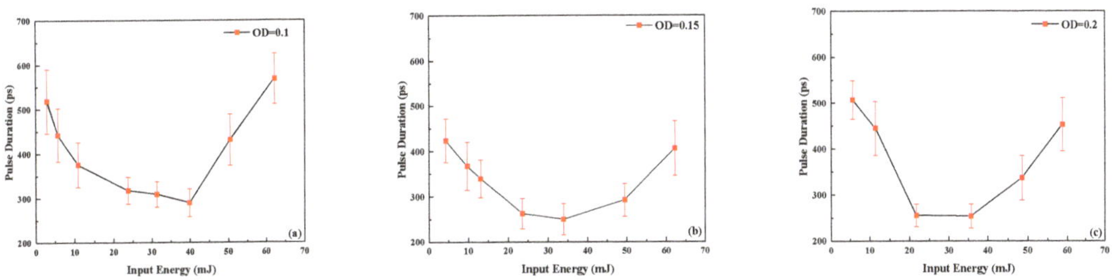

Figure 3. Pulse duration as a function of the input energy after the LIB pulse compression stage: (**a**) OD = 0.1; (**b**) OD = 0.15; (**c**) OD = 0.2.

Taking Figure 3c as an example, when the input energy reaches 21.7 mJ, the output pulse width gradually decreases, and the shortest pulse width is 255.7 ps. When the input energy is increased to 35.7 mJ, the output pulse width remains basically unchanged, and the output pulse width is 254.4 ps. When the input energy continues to be increased to 58.7 mJ, the output pulse width gradually increases, and the longest output pulse width is 453.2 ps.

The reason for this change is that the pulse width of the LIB output is affected by the pulse width of the Stokes beam after SBS compression and due to the LIB switching properties. The pulse width of the Stokes beam compressed by SBS is affected by the output power density of the Nd:YAG pumping source. When the output power density of the Nd:YAG laser is less than a specific value, the pulse width of the Stokes beam rapidly shortens and eventually saturates to a minimum value. The output power density of the Nd:YAG laser is further increased, and the pulse width of the Stokes beam is slowly broadened. In terms of stability, the pulse–width stability is optimal when the output pulse width is the shortest, and the stability gradually deteriorates by further increasing the pump power density [20–23]. By comparing Figure 3a–c, it can be seen that the minimum pulse width of the LIB output and its stability, and the corresponding input energy, are affected by the OD value of LIB, as shown in Table 1.

Table 1. Influence of different OD values on the minimum pulse output by LIB.

Number	OD	Input Energy (mJ)	Minimum Pulse Width (ps)	Stability (ps)
1	0.1	39.9	290.2	31.1
2	0.15	33.9	250	34.4
3	0.2	35.7	254.4	48.6

In the CCl_4 solution doped with the SiC nanowire, the curves of LIB compression output energy and energy conversion efficiency with input energy are shown in Figure 4. Taking Figure 4c as an example, when the input energy is 5 mJ, a stable LIB shutdown phenomenon begins to appear. This is because SiC nanowires can reduce the breakdown threshold of LIB, causing LIB to occur at low energy, significantly suppressing the occurrence of SBS and SRS, making LIB dominant in the competition for nonlinear effects. As the input energy increases, the output energy of LIB gradually increases. Before entering the saturation gain region, the energy conversion efficiency increases rapidly as the input energy rises. When entering the saturation gain region, the energy conversion efficiency gradually tends to saturate and stabilize at about 50% [24–26]. At the same time, the variance in the output energy is also continuously increasing.

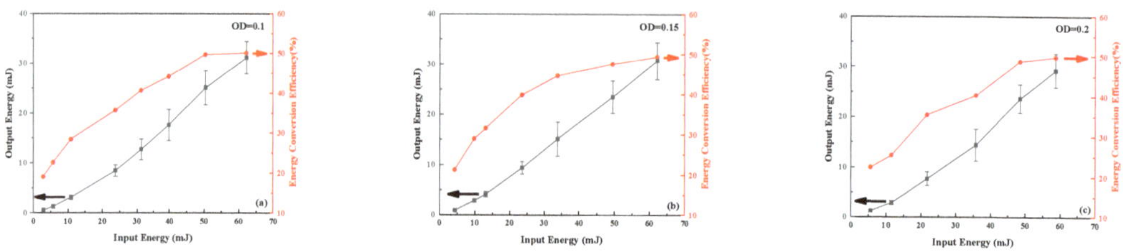

Figure 4. Relationship between output energy and energy conversion efficiency with input energy: (**a**) OD = 0.1; (**b**) OD = 0.15; (**c**) OD = 0.2.

By comparing Figure 4a–c, it can be seen that the output energy and energy conversion efficiency follow the same trend as the input energy. After experiencing a small signal gain under low input energy, the final energy conversion efficiency is saturated at about 50%. This is because the media selected in the LIB pool are the same, resulting in the same phonon lifetime. The change in OD value from 0.1 to 0.2 has a relatively small impact on the gain coefficient. The maximum energy output of LIB and its saturation energy conversion efficiency, the corresponding input energy, are affected by the OD value of the LIB, as shown in Table 2.

Table 2. Influence of different OD values on the output energy by LIB.

Number	OD	Input Energy (mJ)	Maximum Output Energy (mJ)	Saturation Energy Conversion Efficiency (%)
1	0.1	62.2	31.2	50.1
2	0.15	62.2	30.7	49.4
3	0.2	58.7	29.2	49.7

After SBS compression, the output laser energy is 35.7 mJ, and the corresponding energy conversion efficiency and pulse width are 55% and 506.8 ps, respectively. For the LIB pool with an OD value of 0.2, when the input energy is 35.7 mJ, the statistical histogram of LIB output energy stability is shown in Figure 5. The corresponding average value and R^2 value are 12.2 mJ and 92.3%, respectively.

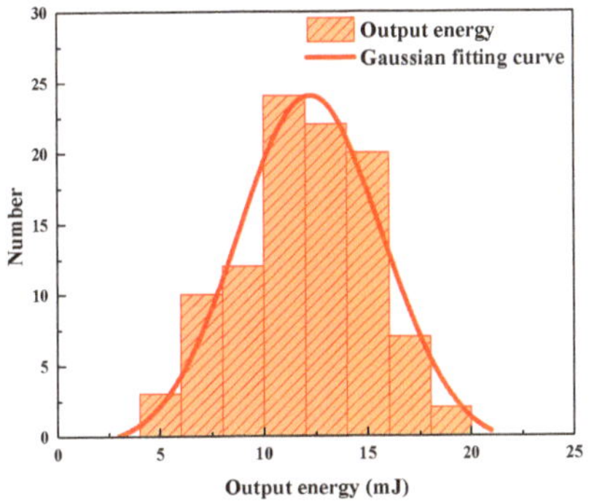

Figure 5. Output energy stability statistical histogram at OD value of 0.2 and input energy of 35.7 mJ.

The comparison of pulse waveforms before and after LIB compression is shown in Figure 6. The black curve represents the pump pulse before LIB compression, and the red curve represents the short pulse after compression. It can be seen that the leading edges of the two pulses are essentially coincident, which indicates that the LIB plasma switch preserves the leading–edge portion of the compression pulse more completely. The trailing edge after the peak is almost absorbed, causing compression. After compression, the slope of the trailing edge of the pulse is basically the same as that of the front edge, forming a short pulse with symmetric front and trailing edges. Due to the effective suppression of other nonlinear effects, the compression effect here is changed from a combination of multiple nonlinear effects in the pure media to a single LIB compression. The final output pulse width is 254.4 ps, and the compression rate of 50.2% is achieved via SBS compression.

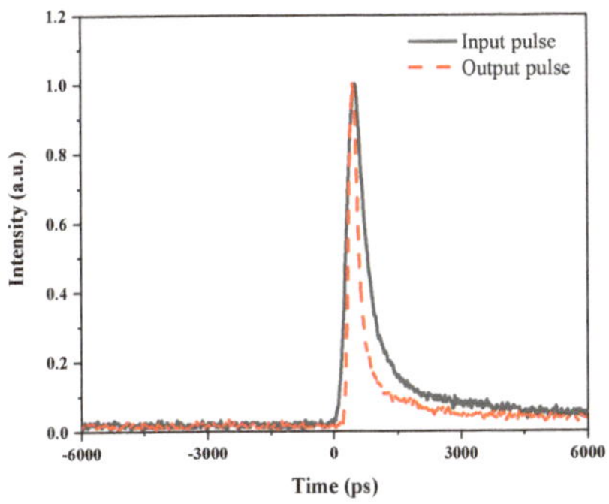

Figure 6. Temporal profiles of unclipped pulse (solid line) and clipped pulse (dash line).

5. Conclusions

In conclusion, we proposed a novel narrow pulse–width compression scheme that combined SBS pulse–width compression technology with LIB switching. The LIB threshold is decreased by doping a SiC nanowire into a CCl_4 solution. The nonlinear effects of SBS and SRS are suppressed to improve the stability of the system. For the LIB switch with an OD value of 0.2, the optimal output is obtained when the input energy is 35.7 mJ. For a LIB switch with an OD value of 0.2, the minimum pulse–width output of 254.4 ps is obtained when the input energy is 35.7 mJ. The corresponding energy conversion efficiency and pulse compression rate are 34.2% and 50.2%, respectively. The produced optical pulse provides a spatial resolution that is well suited for LIDAR Thomson scattering diagnostics.

Author Contributions: Validation, L.F.; writing—original draft preparation, L.F.; writing—review and editing, D.S. and Y.Z.; supervision, L.F., D.S., Y.Z. and W.Z.; project administration, W.Z.; funding acquisition, W.Z. All authors have read and agreed to the published version of the manuscript.

Funding: This research received no external funding.

Institutional Review Board Statement: Not applicable.

Informed Consent Statement: Not applicable.

Data Availability Statement: Data are contained within the article.

Conflicts of Interest: The authors declare no conflicts of interest.

References

1. Ishii, N.; Turi, L.; Yakovlev, V.S.; Fuji, T.; Krausz, F.; Baltuška, A.; Butkus, R.; Veitas, G.; Smilgevičius, V.; Danielius, R.; et al. Multimillijoule chirped parametric amplification of few–cycle pulses. *Opt. Lett.* **2000**, *30*, 567–569. [CrossRef]
2. Bertolotti, M. High–order Harmonic Generation in Laser Plasma Plumes by Rashid Ganeev. *Contemp. Phys.* **2015**, *56*, 88–89.
3. Walsh, M.J.; Beurskens, M.; Carolan, P.G.; Gilbert, M.; Loughlin, M.; Morris, A.W.; Riccardo, V.; Xue, Y.; Huxford, R.B.; Walker, C.I. Design challenges and analysis of the iter core lidar thomson scattering system. Review of Scientific Instruments. *Rev. Sci. Instrum.* **2006**, *77*, 1925. [CrossRef]
4. Au, J.A.D.; Spühler, G.J.; Südmeyer, T.; Paschotta, R.; Hövel, R.; Moser, M.; Erhard, S.; Karszewski, M.; Giesen, A.; Keller, U. 16.2–W average power from a diode–pumped femtosecond Yb:YAG thin disk laser. *Opt. Lett.* **2000**, *25*, 859–861.
5. Liu, Z.; Izumida, S.; Ono, S.; Ohtake, H.; Sarukura, N. High–repetition–rate, high–average–power, mode–locked Ti:sapphire laser with an intracavity continuous–wave amplification scheme. *Appl. Phys. Lett.* **1999**, *74*, 3622–3623.
6. Beddard, T.; Sibbett, W.; Reid, D.T.; Garduno-Mejia, J.; Jamasbi, N.; Mohebi, M. High–average–power, 1–MW peak–power self–mode–locked Ti: Sapphire oscillator. *Opt. Lett.* **1999**, *24*, 163–165. [CrossRef]
7. Zayhowski, J.J.; Dill, I.C. Diode–pumped passively Q–switched picosecond microchip lasers. *Opt. Lett.* **1994**, *19*, 1427–1429. [CrossRef]
8. Braun, B.; Kärtner, F.X.; Zhang, G.; Moser, M.; Keller, U. 56–ps passively Q–switched diode–pumped microchip laser. *Opt. Lett.* **1997**, *22*, 381–383.
9. Kulagin, O.V.; Gorbunov, I.A.; Sergeev, A.M.; Valley, M. Picosecond Raman Compression Laser at 1530 nm with Aberration Compensation. *Opt. Lett.* **2013**, *38*, 3237–3240. [PubMed]
10. Kubeček, V.; Hamal, K.; Procházka, I.; Valach, P.; Buzelis, R.; Dementev, A. Compression of the Nd: YAP Laser Pulse by Two–Stage Stimulated Backward Scattering. *Opt. Commun.* **1989**, *73*, 251–256. [CrossRef]
11. Kuwahara, K.; Takahashi, E.; Matsumoto, Y.; Kato, S.; Owadano, Y. Short–Pulse Generation by Saturated KrF Laser Amplification of a Steep Stokes Pulse Produced by TwoStep Stimulated Brillouin Scattering. *J. Opt. Soc. Am. B* **2000**, *17*, 1943–1947. [CrossRef]
12. Hasi, W.L.J.; Lu, Z.W.; Lu, H.H.; Fu, M.L.; Gong, S.; Lin, D.Y.; He, W.M.; Gao, W. Investigation on Pulse Compression Based on Stimulated Brillouin Scattering and Optical Breakdown. *Appl. Phys. B* **2010**, *98*, 397–400. [CrossRef]
13. Liu, Z.; Wang, Y.; Wang, H.; Bai, Z.; Li, S.; Zhang, H.; Wang, Y.; He, W.; Lin, D.; Lu, Z. Pulse temporal compression by two–stage stimulated Brillouin scattering and laser–induced breakdown. *Appl. Phys. Lett.* **2017**, *110*, 241108.
14. Noack, J.; Vogel, A. Laser–Induced Plasma Formation in Water at Nanosecond to Femtosecond Time Scales: Calculation of Thresholds, Absorption Coefficients, and Energy Density. *IEEE J. Quantum Electron.* **1999**, *35*, 1156–1167. [CrossRef]
15. Bhatnagar, M.; Baliga, B.J. Comparison of 6H–SiC, 3C–SiC, and Si for power devices. *IEEE Trans. Electron Devices* **1993**, *40*, 645–655. [CrossRef]
16. Shelton, D.P. Long–range correlation of intra–molecular and inter–molecular vibration in liquid CCl_4. *J. Chem. Phys.* **2021**, *154*, 034502. [CrossRef]

17. Chakraborty, T.; Rai, S.N. Depolarization ratio and correlation between the relative intensity data and the abundance ratio of various isotopes of liquid carbon tetrachloride at room temperature. *Spectrochim. Acta A Mol. Biomol. Spectrosc.* **2005**, *62*, 438–445. [CrossRef] [PubMed]
18. Kuwahara, K.; Takahashib, E.; Matsumoto, Y.; Matsushima, I.; Okuda, I.; Kato, S.; Owadano, Y. High Intensity Pulse Generation by Saturated Amplification of Stokes Pulse with Steep Leading Edge. In Proceedings of the ECLIM 2000: 26th European Conference on Laser Interaction with Matter, Prague, Czech Republic, 12–16 June 2000; SPIE: Bellingham, WA, USA, 2011; Volume 4424, pp. 155–158.
19. Mitra, A.; Yoshida, H.; Fujita, H.; Nakatsuka, M. Sub Nanosecond Pulse Generation by Stimulated Brillouin Scattering Using FC–75 in an Integrated with Laser Energy up to 1.5 J. *Jpn. J. Appl. Phys.* **2006**, *45*, 1607–1611.
20. Feng, C.; Xu, X.; Diels, J.C. High–Energy Sub–Phonon Lifetime Pulse Compression by Stimulated Brillouin Scattering in Liquids. *Opt. Express* **2017**, *25*, 12421–12434. [CrossRef]
21. Frings, H. Compact Temporal–Pulse–Compressor Used in Fused–Silica Glass at 1064 nm Wavelength. *Jpn. J. Appl. Phys.* **2007**, *46*, L80–L82.
22. Dane, C.B.; Neuman, W.A.; Norton, M.A. Energy Scaling of SBS Pulse Compression. *Proc. SPIE* **1992**, *1626*, 297–307.
23. Hall, T.J. Principles of Phase Conjugation. *Opt. Acta Int. J. Opt.* **1986**, *33*, 685–686. [CrossRef]
24. Yuan, H.; Wang, Y.; Lu, Z.; Wang, Y.; Liu, Z.; Bai, Z.; Cui, C.; Liu, R.; Zhang, H.; Hasi, W. Fluctuation Initiation of Stokes Signal and Its Effect on Stimulated Brillouin Scattering Pulse Compression. *Opt. Express* **2017**, *25*, 14378–14388.
25. Hasi, W.; Zhong, Z.; Qiao, Z.; Guo, X.; Li, X.; Lin, D.; He, W.; Fan, R.; Lü, Z. The Effects of Medium Phonon Lifetime on Pulse Compression Ratio in the Process of Stimulated Brillouin Scattering. *Opt. Commun.* **2012**, *285*, 3541–3544.
26. Gorbunov, V. Formation and Amplification of Ultrashort Optical Pulses as a Result of Stimulated Scattering in Opposite Directions. *Sov. J. Quantum Electron.* **1984**, *14*, 1066–1069.

photonics

MDPI

Article

Exploration of Illicit Drug Detection Based on Goos–Hänchen Shift

Yan Wang [1], Xiaodi Zhou [2], Xinmin Fan [1,*], Xiaodong Huang [1], Lujun Zhang [1] and Chunyan Wang [1,*]

[1] School of Physics and Electronic Information, Weifang University, Weifang 261061, China; wfwy90@wfu.edu.cn (Y.W.); xdhuang@wfu.edu.cn (X.H.); zhang_lujun1985@wfu.edu.cn (L.Z.)

[2] Research and Development Centre, Shandong North Optical Electronics Co., Ltd., Taian 271000, China; zhouxiaodistudy@163.com

* Correspondence: xinminfan@wfu.edu.cn (X.F.); wangcy@bnu.edu.cn (C.W.)

Abstract: Amidst the escalating issue of drug abuse, an urgent need for effective illicit drug detection methods has arisen. This paper introduces a novel optical approach utilizing the Goos–Hänchen Shift (GHS) to explore the possibility of on-site rapid detection of illicit drugs. Delving into the mechanisms, light absorption and attenuation in biological samples are considered through absorption and attenuation coefficients, establishing connections between complex refractive indices, complex dielectric constants, and GHS. A self-assembled GHS detection system measured GHS values across various samples: ultrapure water, serum, methamphetamine, serum–methamphetamine, heroin, and serum–heroin. These experiments unveiled substantial GHS variations among the samples. Refractive indices for serum, serum–methamphetamine, and serum–heroin samples were computed using GHS values and sample extinction coefficients, highlighting GHS's remarkable sensitivity to refractive index variations as a high-sensitivity refractive index sensing technology. The correlation between the dielectric constant and GHS was explored, yielding refractive indices for pure solutes—serum, methamphetamine, and heroin—of 1.66300, 1.51300, and 1.62300, respectively. Notably, the dielectric constants for these solutes were 2.76557, 2.28917, and 2.63413, emphasizing the dielectric constant's discriminative potential in identifying illicit drugs. In conclusion, these findings suggest that GHS holds promise for distinguishing various illicit drug types, charting an innovative path for illicit drug detection.

Keywords: Goos–Hänchen Shift; illicit drug detection; refractive index; dielectric constant

1. Introduction

Illicit drugs dependence and addiction pose severe threats to human mental and physical health, familial relationships, and societal stability. This escalating global concern underscores the urgency of combatting drug abuse. Advances in drug detection technologies have significantly bolstered these efforts [1–3]. Current drug detection methods are primarily bifurcated into separation-based detection and direct detection.

Separation-based detection primarily employs techniques such as LC-MS (Liquid Chromatography–Mass Spectrometry) [4–6] and CE (Capillary Electrophoresis) [7]. LC-MS, being adept at both qualitative and quantitative drug analyses from complex body fluids and having commendable stability with a detection limit as low as 10 ng/mL [8], is the gold standard for drug testing. Its results can serve as evidence in the criminal justice system [9]. However, LC-MS requires prolonged analysis times (over 5 h), skilled technicians, and costly equipment. Its lab-bound nature renders it unsuitable for rapid drug testing needs [10,11]. CE, while combining the strengths of various separation techniques and being efficient and automated [12], suffers from reproducibility and stability issues due to its short and narrow analytical pathways [13], preventing it from becoming a mainstream drug detection method.

Direct detection predominantly employs Surface Enhanced Raman Spectroscopy (SERS). Optical technologies, such as SERS, offer advantages like minimal sample contact, no pre-treatment, lesser sample requirements, and swift detection processes [14]. SERS is now widely used for drug detection [15,16] and has almost become the only method for directly detecting drug varieties. However, measurements via SERS must be carried out on precious metal nanoparticle substrates, which have hard-to-control enhancement effects, leading to poor reproducibility. The interference of these metal nanoparticles with biological samples also cannot be overlooked [17]. Thus, there is a pressing need to explore a new, highly sensitive, and accurate direct detection method for advancing illicit drug detection technologies.

The GHS (Goos–Hänchen Shift) denotes the lateral displacement of a reflected light beam from its expected geometrical path during total internal reflection [18]. This shift serves as an apt detection signal, unaffected by variations in light source intensity [19]. Since its experimental discovery in 1947, GHS has spurred numerous theoretical and empirical studies and has found applications in sensors [20–22], optical switches [23,24], and optical storage [25]. Notably, with GHS's tight association with the electromagnetic properties of materials (like refractive index), it is emerging as a detection method in chemical and environmental realms [26,27]. Recent research by Min Jin in 2023, for instance, leveraged a specifically designed GHS sensor for detecting aquatic heavy metal ions [20]. Jiangyu Liu explored the feasibility of using a designed terahertz GHS sensor for detecting biological small molecules [26]. Elnaz demonstrated the efficacy of GHS in detecting low-concentration biomolecules [27]. Due to the small value of GHS, it is relatively difficult to measure, so enhancing the GHS displacement has become a research hotspot. SPR sensors based on GHS [20,28–30] have been widely used in the detection field and have achieved good detection sensitivity. However, just like the SERS technology, the SPR enhancement effect depends on the presence of metal nanoparticles or nanofilms, and the impact of precious metals on biological samples is still unknown. Therefore, for biological samples, we prefer the direct detection method. These findings intimate that GHS, as a sensitive optical direct detection method, might be well-suited for illicit drug detection, which this study sets out to investigate.

In this study, we constructed the GHS measurement system to detect the GHS values of various samples, encompassing ultrapure water, serum, methamphetamine, heroin, serum–methamphetamine, and serum–heroin mixtures. Considering the absorption and scattering of the samples, we established a relationship between the complex refractive index of the samples and GHS using the steady-state phase method GHS calculation formula. From the experimentally measured GHS and extinction coefficients, we derived the refractive index information for serum, serum–methamphetamine, and serum–heroin samples based on the aforementioned relationship. Using the refractive index calculation formula for mixed solutions, we further analyzed the serum, serum–methamphetamine, and serum–heroin samples. This analysis enabled us to determine the refractive indices and dielectric constants of the three pure substances: serum, methamphetamine, and heroin. Furthermore, we proposed a solution method to determine the dielectric constant of unknown illicit drug components in the serum. Our findings offer valuable insights for distinguishing between different types of illicit drugs in drug detection processes.

2. Principles and Methods

2.1. Principle of Goos–Hänchen Shift

When a light beam transitions from an optically dense medium to a less dense one, and the angle of incidence surpasses the critical angle for total internal reflection, full reflection occurs at the interface. In classical geometric optics, it is conventionally assumed that the phenomenon of total internal reflection adheres to the law of geometric reflection, implying that the incident and reflected rays coincide at the same position. However, this is not entirely accurate. Goos and Hänchen [18], through experimentation, observed that under conditions of total internal reflection, the reflected light beam deviates from the ideal geometric path by a small distance, denoted as the Goos–Hänchen Shift, as shown in Figure 1.

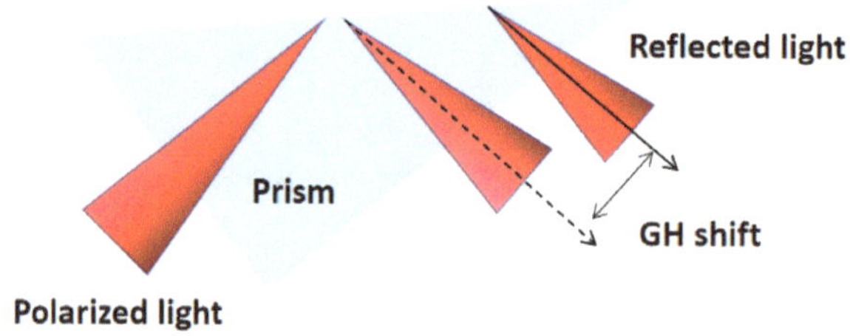

Figure 1. Schematic representation of the Goos–Hänchen Shift under total internal reflection conditions. The solid line represents the actual reflected light beam, while the dashed line represents the geometrically expected reflected light beam.

Currently, GHS is primarily studied using the steady-state phase method. The incident light can be considered as a superposition of monochromatic plane waves with different wavevector directions. Consequently, the full reflected beam is composed of various corresponding reflected plane waves, each having different phases, leading to the displacement of the reflected light. TM-polarized light exhibits larger GHS values compared to TE-polarized light. GHS measurements are carried out using polarization-based methods, specifically by measuring the displacement difference in GHS between TE-polarized and TM-polarized light [18]. By subtracting the results obtained under TM and TE conditions, the influence of small measurement errors on the test results can be mitigated, thereby improving the accuracy of the system's measurements.

2.2. *Experimental Device*

A GHS measurement system was constructed, as shown in Figure 2. The entire system mainly consists of polarization modulation, incident angle control, displacement generation, and detection components. Parallel beams of light, collimated by a lens assembly, pass through the first polarizer, converting the beams into linearly polarized light (with a polarization direction at a 45° angle to the horizontal plane). Then, the linearly polarized light goes through the second adjustable polarizer to modulate the polarization state of the light into either TE or TM polarization. In this study, a prism was used as the displacement-generating component. The light beam enters the prism from one side of its right-angled surface, and when it contacts the sample interface in the prism's sloping surface, total internal reflection occurs, resulting in the generation of the GHS. The reflected light exits from the other right-angled side of the prism and is directed onto a detector. The prism is mounted on a rotation stage, and the incident angle of the light beam for total internal reflection can be changed by automatically controlling the rotation stage. The system operates as follows:

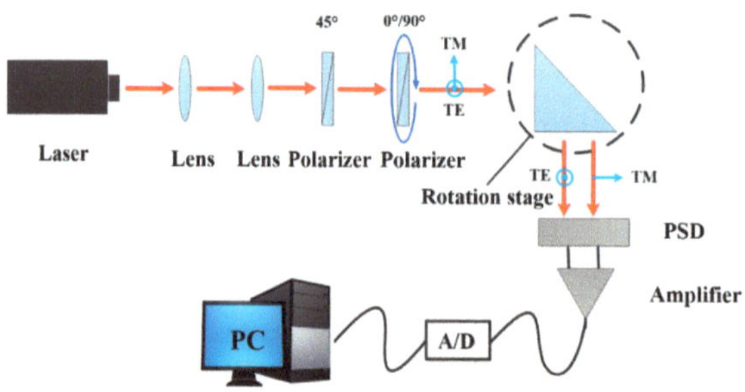

Figure 2. Schematic diagram of GHS measurement system.

A He-Ne laser (1107P, JDSU, Milpitas, CA, USA) emitting light with a wavelength of 632.8 nm produces a collimated light beam with a spot diameter of 1 mm after passing through a lens assembly. The first polarizer (Edmund Optics, Barrington, IL, USA) is used to adjust the polarization angle to 45°, resulting in linearly polarized light. Then, the second polarizer is rotated to adjust the polarization direction of the incident light. When the transmission direction of the second polarizer is set to be horizontal (0°) or vertical (90°), TM- and TE-polarized light are obtained, respectively. Furthermore, the light enters an isosceles right-angle prism (PS911, Thorlabs, Newton, NJ, USA), where it undergoes reflection at the prism's sloping surface. The reflected light's position signal, detected by a position-sensitive detector (PSD, ON-TRAK Optics, Milpitas, CA, USA), is amplified through an amplifier (OT-301, ON-TRAK Optics, Milpitas, CA, USA), digitized through A/D conversion, and then input into a computer for data processing.

2.3. Experimental Samples

The methamphetamine and heroin used in this experiment were research-grade and were obtained from the Narcotics Detection Center of the Tianjin Public Security Bureau. The serum (fetal bovine serum, F4135) was purchased from Sigma, Kenilworth, NJ, USA, and all samples were used without further purification. Ultrapure water was used as the solvent, and solutions of serum, methamphetamine, and heroin were prepared with a concentration of 10^{-3} mol/L. Mixtures of methamphetamine and heroin with serum were also prepared to create serum–methamphetamine and serum–heroin mixed solutions. All experimental images presented in this article represent the average results of five repeated experiments, and there are no significant differences between the results of each repetition.

3. Results and Discussion

To ensure the accuracy of the experimental results, the errors in the GHS measurement system shown in Figure 2 were first analyzed. During the experiment, the light beams were adjusted to either the TM or TE polarization, and the TM and TE polarizations might not have aligned perfectly, resulting in a certain distance between them. Multiple measurements revealed that the distance between them was approximately 2 μm. To eliminate the influence of this distance on GHS measurements, it was necessary to calibrate the zero point of the GHS, i.e., correcting for this factor.

As shown in Figure 3, during the adjustment of the polarizer, there was a distance "L" between the TM and TE polarizations. After passing through the prism, if total internal reflection did not occur, the distance between the two polarizations remained "L." However, if total internal reflection occurred, the distance between the polarizations became "L + S2-S1," where "S1" and "S2" represent the GHS values for the TE and TM polarizations, respectively. Therefore, by subtracting "L" from the measured difference between the positions of the two polarizations when total internal reflection occurred (L + S2-S1), the actual relative displacement value could be obtained as "S2-S1." When total internal reflection did not occur, both the TM and TE polarizations had GHS values of 0, resulting in a GHS value of 0. Thus, the correction value "L" could be obtained by measuring the distance between the different polarization states when total internal reflection did not occur. This correction allowed us to account for the influence of angular changes on GHS measurements by differentiating between the positions of the reflected light beams under different polarizations, effectively preserving only the GHS value.

The quality of reflected light spots at different incident angles was analyzed. Three arbitrary wavelengths of TE-polarized and TM-polarized light, each with a wavelength of 632.8 nm, were incident on the prism–air interface, and the reflected light spots were processed using MATLAB R2020b software. This processing involved obtaining the intensity distribution along a horizontal line passing through the center of the light spot, as shown in Figure 4. During the image processing, the light spots were translated to align their contours to enable a clearer observation of any changes in the light spot's shape. The results indicate that the shape of the light spots remains essentially unchanged at different incident angles, with energy concentrated at the center of the light spot, and any impact on the beam position detection can be neglected.

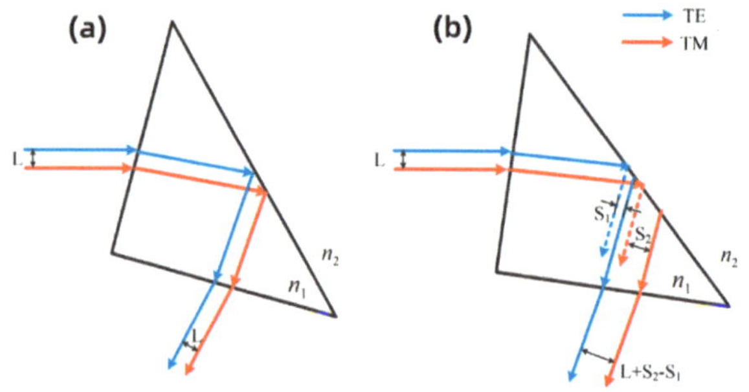

Figure 3. Polarizer adjustment leading to beam displacement. (**a**) No total reflection occurred; (**b**) total reflection occurs.

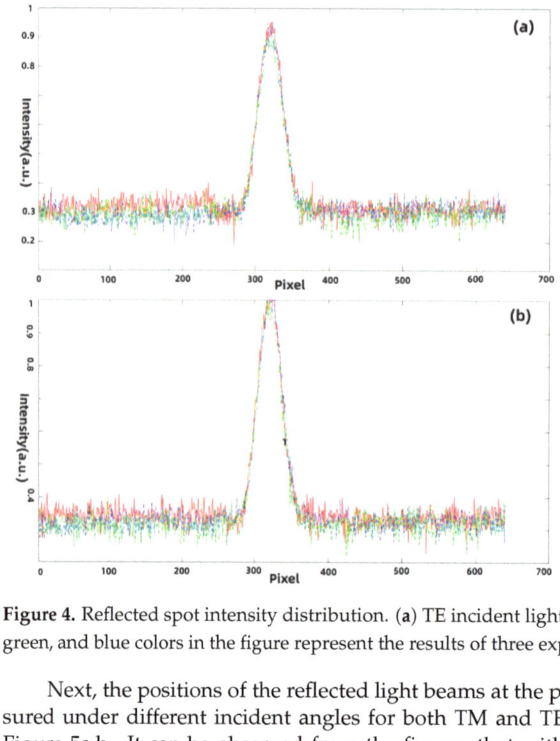

Figure 4. Reflected spot intensity distribution. (**a**) TE incident light, (**b**) TM incident light. The red, green, and blue colors in the figure represent the results of three experiments.

Next, the positions of the reflected light beams at the prism–air interface were measured under different incident angles for both TM and TE polarizations, as shown in Figure 5a,b. It can be observed from the figures that with an increase in the incident angle, the light spot moves in the positive direction of the detector's X-axis. During the experimental measurements, the distance between the TM polarization and the TE polarization when no total internal reflection occurred was measured to be 1.3 μm. By taking the difference in the positions of the light beams between Figure 5a,b and subtracting the correction value of 1.3 μm, the position values of the prism–air interface (GH position) were obtained, as shown in Figure 5c. The prism's refractive index used in the experiment was 1.515 (laser wavelength: 632.8 nm), and the refractive index of air was 1. Based on calculations, the critical angle for total internal reflection was determined to be 41.30°. To reduce experimental errors, multiple sets of data were measured and averaged, and efforts were made to ensure that the light beam was incident vertically on the PSD photosensitive surface. Near the critical angle for total internal reflection, the measured displacement peak was 7.16 μm, which is approximately 11.3 times the wavelength. As the incident angle increased, the displacement rapidly decreased and then stabilized at around 2 μm. Furthermore, according to the steady-state phase method theory, the theoretical values of the GHS for the prism–air interface were calculated, as shown by the red solid line in Figure 5c. It can be observed that the theoretical values closely matched the experimental measurements, indicating the accuracy of the measurement setup built in this study.

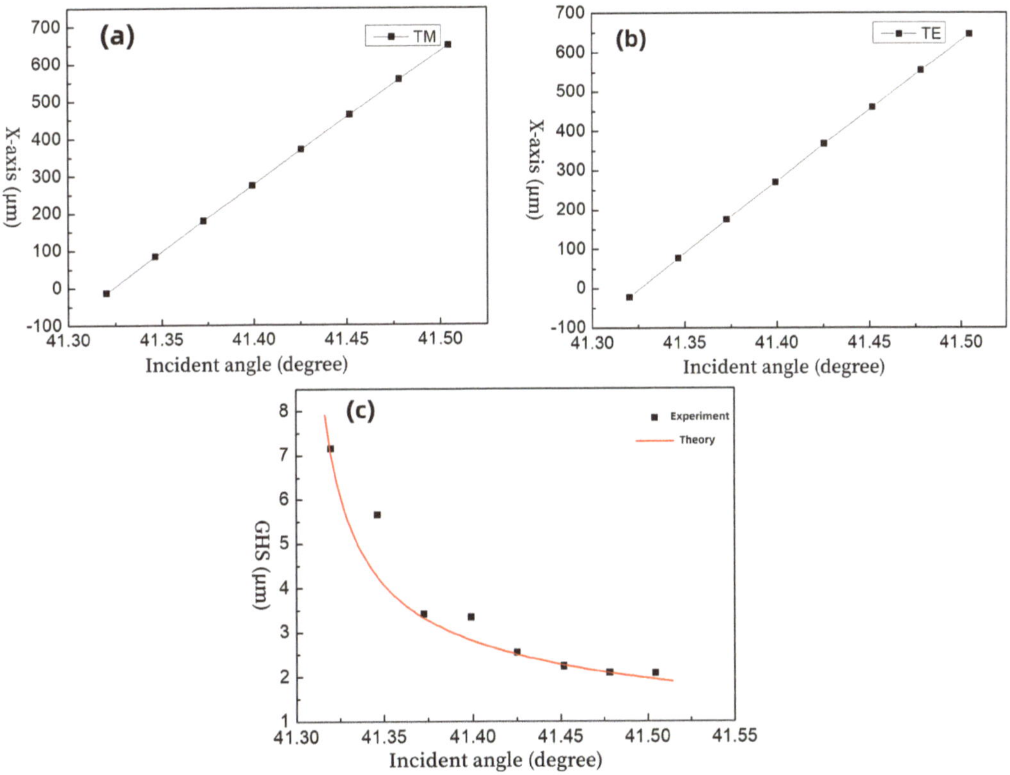

Figure 5. (**a**) Distribution of TM-polarized reflected light positions at the prism–air interface, (**b**) distribution of TE-polarized reflected light positions at the prism–air interface, and (**c**) comparison of experimental and theoretical GHS values at the prism–air interface.

In contrast to the refractive index, the concept of the complex refractive index encompasses fundamental optical parameters such as the refractive index and the absorption coefficient. It stands as a critical physical quantity for characterizing the optical properties of substances and holds significant relevance as an optical parameter for biological samples [31]. Notably, some prior studies, as documented in references [26,27], did not account for the impact of physical phenomena like absorption and attenuation within samples on the refractive index. However, biological samples, typified by serum, exhibit light absorption or attenuation within the ultraviolet to visible wavelength spectrum. These optical behaviors have the potential to influence the characteristics of reflected light and, by extension, the GHS. Consequently, considering the attenuation and absorption of substances in response to light [32–34], we conducted thorough derivations and analyses to elucidate the intricate relationship between the GHS and the complex refractive index of materials.

The important formulas are displayed as follows. Please refer to the Supplementary document for detailed formula derivation.

It can be seen that the polarization rate of the medium is complex, therefore, the refractive index of the material can be expressed as a complex number:

$$\hat{n} \approx \sqrt{\varepsilon_r} = \sqrt{1+\chi} = n + i\eta \tag{1}$$

In the formula, n is the refractive index of the medium, ε_r is the dielectric constant, and η is the extinction coefficient of the medium. Furthermore, the complex refractive index can be related to the complex dielectric function:

$$\hat{\varepsilon}_r = 1+\chi = n^2 - \eta^2 + 2n\eta i \tag{2}$$

After rigorous deduction, we can calculate the phase of the reflection coefficients as

$$\phi_s = \mathrm{Im}\left\{ \ln\left[\frac{n_1\cos\theta_1 - \sqrt{\left(n_2 + i\frac{(\mu_a+\mu_s)\lambda}{4\pi}\right)^2 - n_1^2\sin^2\theta_1}}{n_1\cos\theta_1 + \sqrt{\left(n_2 + i\frac{(\mu_a+\mu_s)\lambda}{4\pi}\right)^2 - n_1^2\sin^2\theta_1}} \right] \right\} \tag{3}$$

$$\phi_p = \mathrm{Im}\left\{ \ln\left[\frac{\left(n_2 + i\frac{(\mu_a+\mu_s)\lambda}{4\pi}\right)^2\cos\theta_1 - \sqrt{n_1^2\left(n_2 + i\frac{(\mu_a+\mu_s)\lambda}{4\pi}\right)^2 - n_1^2\sin^2\theta_1)}}{\left(n_2 + i\frac{(\mu_a+\mu_s)\lambda}{4\pi}\right)^2\cos\theta_1 + \sqrt{n_1^2(\left(n_2 + i\frac{(\mu_a+\mu_s)\lambda}{4\pi}\right)^2 - n_1^2\sin^2\theta_1)}} \right] \right\} \tag{4}$$

with these phase values, the GHS can be expressed as

$$D_i = -\frac{\lambda_0}{2\pi n_1}\frac{d\phi_i}{d\theta_1} = -\frac{\lambda_0}{2\pi n_1}\frac{d\mathrm{Im}[\ln(\phi_i)]}{d\theta_1} \tag{5}$$

It can be observed that the magnitude of the GHS is influenced by several factors, including the wavelength of light, polarization state, angle of incidence, and refractive index of the medium, as well as the absorption and scattering coefficients. However, when the wavelength of light, polarization state, and angle of incidence are held constant, the magnitude of the GHS is determined by the refractive index, absorption coefficient, and scattering coefficient, which reflect the electromagnetic properties of the material itself. In essence, the GHS can serve as a characterization of the intrinsic electromagnetic properties of the material.

Currently, illicit drugs are primarily categorized as new drugs and traditional drugs. Methamphetamine represents a typical new drug, often referred to as synthetic drugs, and it is also one of the most widely abused drugs worldwide. Heroin, on the other hand, is a classic example of a traditional drug and accounts for a significant proportion of seized drugs. In the following sections, we will explore the characteristics of these two drugs, methamphetamine and heroin, using a GHS analysis.

Different biological samples were subjected to GHS testing and analysis, employing a 632.8 nm laser source. Experimental measurements of GHS values were conducted on six distinct samples: ultrapure water, serum, methamphetamine, heroin, serum–methamphetamine, and serum–heroin. The resulting GHS data are presented in Figure 6, with a summarized overview available in Table 1. The results clearly illustrate significant variations among these samples, indicating the potential of the GHS to distinguish between them.

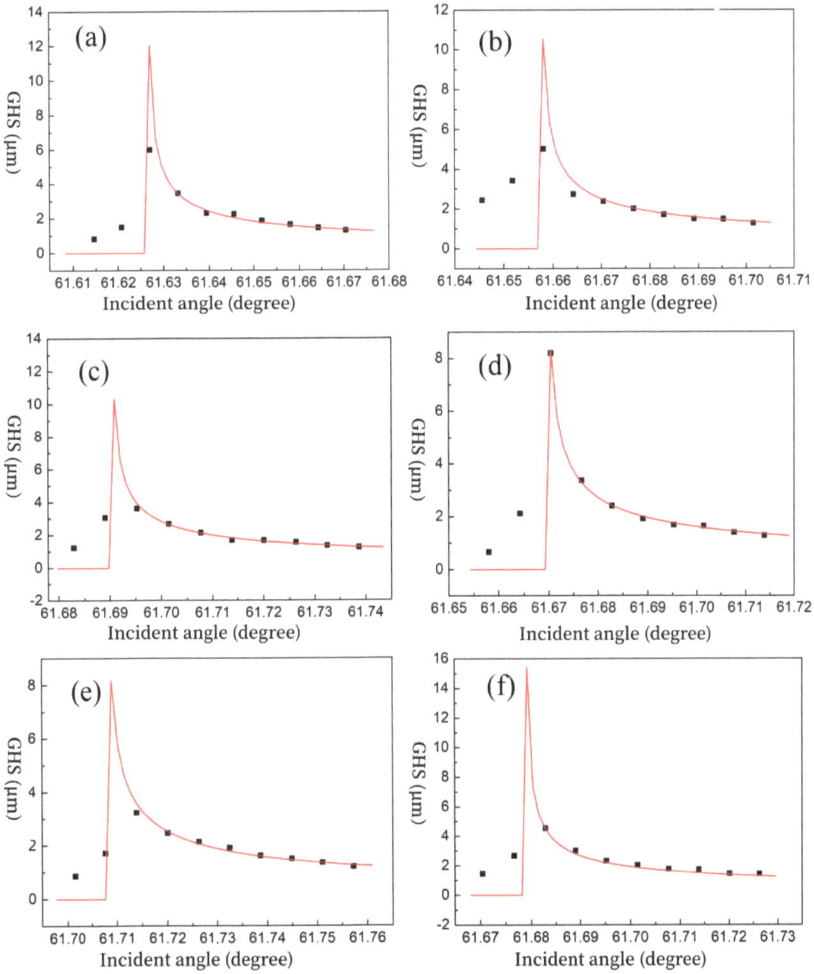

Figure 6. Theoretical fitting and experimental GHS data of samples. (**a**) Ultrapure water, (**b**) serum solution, (**c**) methamphetamine solution, (**d**) serum containing methamphetamine, (**e**) heroin solution, and (**f**) serum containing heroin.

Table 1. GHS values for samples.

Sample	Water	Serum	Methamphetamine	Serum–Methamphetamine	Heroin	Serum–Heroin
Incident angle (degree)	61.627	61.658	61.695	61.670	61.714	61.683
GH displacement (μm)	6.00	5.02	3.62	8.19	3.20	4.53

Furthermore, a theoretical fitting of the GHS values for these samples is depicted by the red solid line in Figure 6. It is evident that the theoretical model closely aligns with the experimental results, particularly beyond the critical angle for total internal reflection. Notably, the GHS patterns for all samples exhibit consistency, featuring displacement peaks in proximity to the critical angle for total internal reflection, followed by a rapid decline in displacement beyond this critical angle.

The differences in the GHS reflect variations in the refractive indices of the test samples. Based on the relationship between the GHS and the complex refractive index of substances (Equations (3)–(5)), we used serum, serum–methamphetamine, and serum–heroin as examples to analyze the refractive indices of serum, methamphetamine, and heroin (pure substances). We also discussed their potential applications in substance differentiation in the field of drug detection. As shows in Figure 6, the positions of the displacement peaks for serum, serum–methamphetamine, and serum–heroin samples were different. To ensure that total internal reflection occurred and to maximize differentiation, the incident angle was selected as 61.670°. The GHS is related to the absorption and scattering coefficients of the samples, so the total attenuation coefficient needed to be measured in the experiment. We represented the total attenuation coefficient as follows:

$$\mu = -\ln\left(\frac{I_t}{I_0 - I_r}\right)/x \tag{6}$$

where

$$I_0 = I_r + I_t + I_\alpha + I_s \tag{7}$$

In the formula, I_0, I_r, I_t, I_α, and I_s represent the intensity of incident, reflected, transmitted, absorbed, and scattered light, respectively. The samples, including serum, serum–methamphetamine, and serum–heroin, were introduced into the flow cell through the sample inlet. The GHS for each sample was measured, and the incident light intensity, reflected light intensity, and transmitted light intensity were measured using a power meter (S120VC, Thorlabs, Newton, NJ, USA). With these measurements, the attenuation coefficients and extinction coefficients of the samples were calculated, respectively. Furthermore, by substituting the calculated extinction coefficients and the measured GHS values into Equations (3)–(6), the refractive indices of the samples were determined, as shown in Table 2. Overall, based on Formulas (3)–(5), by incorporating the parameters from Table 2 into the formula, the reflective index is calculated as a physical quantity. The results indicate that the extinction coefficients and refractive indices of the three samples are relatively close, primarily due to the low concentrations of the samples. With such small differences in refractive indices, it is challenging to differentiate them using other measurement methods, whereas the GHS exhibits significant numerical variations. This indirectly indicates that the GHS is highly sensitive and can serve as a high-sensitivity sensing technique.

Table 2. Summary table of experimental results.

Sample	Serum	Serum–Methamphetamine	Serum–Heroin
GHS/μm	1.70	2.40	4.53
Incident light power/μW	95.3	95.3	95.3
Reflected light power/μW	71.3	70.1	64.8
Transmitted light power/μW	0.43	0.43	0.16
Absorption coefficient/cm^{-1}	5.90	5.97	7.70
Extinction coefficient	2.97×10^{-5}	3.01×10^{-5}	3.88×10^{-5}
Refractive index	1.33333	1.33351	1.33362

If the GHS measurement is to be developed as a detection technique, it is necessary to find a correlation between the intrinsic physical properties of the sample and the GHS. The dielectric constant is a physical property that represents the polarization ability of a dielectric material under the action of an electric field and is an intrinsic physical property. Studies have shown a correlation between the dielectric constant and the refractive index. In the following section, we aim to establish a correlation between the GHS and the dielectric constant of the test sample. In the previous sections, we obtained the GHS, refractive indices, and extinction coefficients for serum, serum–methamphetamine, and serum–heroin samples. However, it is evident that the main component in these samples is water. Next, we will further analyze the samples by excluding water and focusing on the solutes, i.e., the effects of serum, methamphetamine, and heroin (pure substances).

The Arago–Biot formula [35] is one of the theoretical formulas applied to calculate the refractive index of mixed solutions. Taking a two-component mixed solution as an example, the formula is as follows:

$$n_m = (1 - \varphi)n_a + \varphi n_b \tag{8}$$

Above, n_m is the refractive index of the mixed solution, n_a and n_b are the refractive index of the solution component, and φ is the volume fraction of the component. In this study, ultrapure water is employed as the solvent for the serum sample, and the serum molecules constitute the solute. For the serum–methamphetamine sample, ultrapure water serves as the solvent, and a mixture of serum molecules and methamphetamine molecules constitutes the solute. Similarly, for the serum–heroin sample, ultrapure water functions as the solvent, and the solute consists of a mixture of serum molecules and heroin molecules. The concentrations of these samples are pre-established (all at mol/L). Utilizing Equation (8), the refractive indices of serum, methamphetamine, and heroin are calculated as 1.66300, 1.51300, and 1.62300, respectively. Given that these three substances exhibit weak light absorption at 632.8 nm, their extinction coefficients are deemed negligible. Accordingly, the dielectric constants of serum, methamphetamine, and heroin are computed as 2.76557, 2.28917, and 2.63413, respectively, in accordance with Equation (2). This indicates that, compared to refractive indices, the variations in the derived dielectric constants, as an intrinsic parameter, are more pronounced. Using dielectric constants as a parameter to differentiate between sample types offers greater distinctiveness.

In the previous section, we discussed the method of calculating the refractive index and dielectric constant of pure samples such as methamphetamine, serum, and heroin based on experimental results. However, for actual serum samples, the composition and corresponding volume fractions are unknown. Therefore, it is not possible to directly determine the refractive index of the drugs and differentiate them based on GHS values and the mixed solution refractive index formula. We propose an approach to address this problem. Based on the relationship between the GHS of the sample and the complex refractive index, we can obtain the refractive index value for the entire sample. Let us assume that the known volume ratio of water to serum in a standard serum sample is $V_1 : V_2$.

For a tested serum sample containing drugs, its refractive index can be expressed as

$$n_{s1} = \frac{V_1}{V_1 + V_2 + V_3} n_1 + \frac{V_2}{V_1 + V_2 + V_3} n_2 + \frac{V_3}{V_1 + V_2 + V_3} n_3 \tag{9}$$

Among them, n_1, n_2, n_3 are the refractive indices of water, serum, and drugs, and V_1, V_2, and V_3 are the volumes of water, serum, and drugs, respectively. The unknowns in Equation (9) are n_3 and V_3. We need to prepare a new sample by taking a small amount of the tested serum sample containing drugs and adding an equal volume of the standard serum (other volume ratios can also be used, but the volume ratio must be known). Then, the refractive index of the new sample can be expressed as

$$n_{s2} = \frac{2V_1 + \frac{V_1 V_3}{V_1 + V_2}}{2(V_1 + V_2 + V_3)} n_1 + \frac{2V_2 + \frac{V_2 V_3}{V_1 + V_2}}{2(V_1 + V_2 + V_3)} n_2 + \frac{V_3}{2(V_1 + V_2 + V_3)} n_3 \tag{10}$$

By solving Equations (9) and (10), we can obtain V_s and V_d. Subsequently, we can determine the dielectric constant of the unknown drug, differentiating drugs based on their intrinsic properties. The research findings provide a direction for the subsequent exploration of GHS detection in mixed drug samples.

4. Conclusions

Employing the principles of polarization, we have independently engineered a GHS measurement system and assessed its measurement accuracy. Building on this foundation, we measured the GHS values for various samples, including ultrapure water, serum, methamphetamine, serum–methamphetamine, heroin, and serum–heroin, yielding values of 6.00 μm, 5.02 μm, 3.62 μm, 8.19 μm, 3.2 μm, and 4.53 μm, respectively. These results demonstrate the potential of displacement values for distinguishing between different sample types. The introduction of absorption and attenuation coefficients allowed for the establishment of a model linking complex refractive indices with GHS. Based on experimentally measured GHS values and sample extinction coefficients, refractive index values were calculated for serum, serum–methamphetamine, and serum–heroin samples. These findings highlight the remarkable sensitivity of GHS to changes in the refractive index, positioning it as a high-sensitivity refractive index sensing technology. The correlation between the intrinsic physical property, dielectric constant, of the test samples and the GHS was also explored. Calculations based on the formula for determining refractive indices in mixed solutions revealed refractive indices for solutes in the samples, namely serum, methamphetamine, and heroin, to be 1.66300, 1.51300, and 1.62300, respectively. Moreover, the dielectric constants for these solutes were found to exhibit more pronounced differences, at 2.76557, 2.28917, and 2.63413, respectively. These results underscore the potential of the dielectric constant as a more distinguishing parameter for differentiating between illicit drugs, offering a promising avenue for advancing the field of illicit drug detection. In the subsequent work, we will explore drug detection with different concentrations in large sample sizes and various mixed sample types, aiming to further advance and refine GHS drug detection technology.

Supplementary Materials: The following supporting information can be downloaded at: https://www.mdpi.com/article/10.3390/photonics10111270/s1.

Author Contributions: Y.W.: Methodology, Formal analysis, Writing—review and editing, and Data curation. X.Z.: Investigation and Software. X.F.: Funding acquisition and Data curation. X.H.: Methodology and Funding acquisition. L.Z.: Software and Supervision. C.W.: Software and Writing—review and editing. All authors have read and agreed to the published version of the manuscript.

Funding: This research was funded by the National Natural Science Foundation of China under Grant 61805178 and in part by the Natural Science Foundation of Shandong Provincial under Grant ZR2020QA072.

Institutional Review Board Statement: Not applicable.

Informed Consent Statement: Not applicable.

Data Availability Statement: Data are contained within the article and Supplementary Materials.

Conflicts of Interest: The author Xiaodi Zhou is optical engineer in Research and Development Centre, Shandong North Optical Electronics Co., Ltd. The authors declare no conflict of interest.

References

1. Ju, L.; Lyu, A.; Hao, H.; Shen, W.; Cui, H. Deep Learning-Assisted Three-Dimensional Fluorescence Difference Spectroscopy for Identification and Semiquantification of Illicit Drugs in Biofluids. *Anal. Chem.* **2019**, *91*, 9343–9347. [CrossRef] [PubMed]
2. Griffin, A.; Henry, J.; Kirkbride, K.P.; Painter, B.; Linacre, A. A survey of the effects of common illicit drugs on forensic DNA analysis. *Forensic Sci. Int.* **2022**, *336*, 111314. [CrossRef] [PubMed]
3. Zulyadi, R. Obstacle Factors on Evidence System of Urine Test Result on Drugs Cases in National Narcotics Board of North Sumatera Province, Indonesia. *Bp. Int. Res. Crit. Inst. (BIRCI-J.) Humanit. Soc. Sci.* **2020**, *3*, 1316–1324. [CrossRef]

4. Chen, W.; Li, X.; Huang, H.; Zhu, X.; Jiang, X.; Zhang, Y.; Cen, K.; Zhao, L.; Liu, X.; Qi, S. Comparison of gas chromatography-mass spectrometry and gas chromatography-tandem mass spectrometry with electron ionization for determination of N-nitrosamines in environmental water. *Chemosphere* **2017**, *168*, 1400–1410. [CrossRef] [PubMed]
5. Deventer, K.; Pozo, O.J.; Van Eenoo, P.; Delbeke, F.T. Development of a qualitative liquid chromatography/tandem mass spectrometric method for the detection of narcotics in urine relevant to doping analysis. *Rapid Commun. Mass Spectrom.* **2007**, *21*, 3015–3023. [CrossRef]
6. Fiorentin, T.R.; Logan, B.K.; Martin, D.M.; Browne, T.; Rieders, E.F. Assessment of a portable quadrupole-based gas chromatography mass spectrometry for seized drug analysis. *Forensic Sci. Int.* **2020**, *313*, 110342. [CrossRef]
7. Saar-Reismaa, P.; Brilla, C.A.; Leiman, K.; Kaljurand, M.; Vaher, M.; Kulp, M.; Mazina-Šinkar, J. Use of a newly-developed portable capillary electrophoresis analyser to detect drugs of abuse in oral fluid: A case study. *Talanta* **2020**, *211*, 120662. [CrossRef]
8. Trufelli, H.; Palma, P.; Famiglini, G.; Cappiello, A. An overview of matrix effects in liquid chromatography-mass spectrometry. *Mass Spectrom. Rev.* **2011**, *30*, 491–509. [CrossRef]
9. Mosekiemang, T.T.; Stander, M.A.; de Villiers, A. Ultra-high pressure liquid chromatography coupled to travelling wave ion mobility-time of flight mass spectrometry for the screening of pharmaceutical metabolites in wastewater samples: Application to antiretrovirals. *J. Chromatogr. A* **2021**, *1660*, 462650. [CrossRef]
10. Cîntă-Pînzaru, S.; Peica, N.; Küstner, B.; Schlücker, S.; Schmitt, M.; Frosch, T.; Faber, J.H.; Bringmann, G.; Popp, J. FT-Raman and NIR-SERS characterization of the antimalarial drugs chloroquine and mefloquine and their interaction with hematin. *J. Raman Spectrosc.* **2006**, *37*, 326–334. [CrossRef]
11. Swerdlow, H.; Gesteland, R. Capillary gel electrophoresis for rapid, high resolution DNA sequencing. *Nucleic Acids Res.* **1990**, *18*, 1415–1419. [CrossRef] [PubMed]
12. Xu, Y.; Gao, Y.; Wei, H.; Du, Y.; Wang, E. Field-amplified sample stacking capillary electrophoresis with electrochemiluminescence applied to the determination of illicit drugs on banknotes. *J. Chromatogr. A* **2006**, *1115*, 260–266. [CrossRef] [PubMed]
13. Cialla, D.; März, A.; Böhme, R.; Theil, F.; Weber, K.; Schmitt, M.; Popp, J. Surface-enhanced Raman spectroscopy (SERS): Progress and trends. *Anal. Bioanal. Chem.* **2012**, *403*, 27–54. [CrossRef] [PubMed]
14. Fan, R.; Liu, Z.; Jin, D.; Luo, T.; Li, N.; Li, S.; Wang, Y.; Xia, Y.; Lu, Z. High temporal waveform fidelity stimulated Brillouin scattering phase conjugate mirror using Novec-7500. *Opt. Express* **2023**, *31*, 1878–1887. [CrossRef] [PubMed]
15. Dong, R.; Weng, S.; Yang, L.; Liu, J. Detection and direct readout of drugs in human urine using dynamic surface-enhanced Raman spectroscopy and support vector machines. *Anal. Chem.* **2015**, *87*, 2937–2944. [CrossRef] [PubMed]
16. Weng, S.; Dong, R.; Zhu, Z.; Zhang, D.; Zhao, J.; Huang, L.; Liang, D. Dynamic surface-enhanced Raman spectroscopy and Chemometric methods for fast detection and intelligent identification of methamphetamine and 3, 4-Methylenedioxy methamphetamine in human urine. *Spectrochim. Acta A Mol. Biomol. Spectrosc.* **2018**, *189*, 1–7. [CrossRef]
17. Fu, B.B.; Tian, X.D.; Song, J.J.; Wen, B.Y.; Zhang, Y.J.; Fang, P.P.; Li, J.F. Self-Calibration 3D Hybrid SERS Substrate and Its Application in Quantitative Analysis. *Anal. Chem.* **2022**, *94*, 9578–9585. [CrossRef]
18. Goos, F.; Hänchen, H. Ein neuer und fundamentaler Versuch zur Totalreflexion. *Ann. Phys.* **1947**, *436*, 333–346. [CrossRef]
19. Ullah, Z.; Ahmad, S.; Khan, T.; Jan, M.N.; Abdul Jabar, M.S. Complex conductivity dependent Goos–Hanchen shifts through metallic surface. *J. Phys. B At. Mol. Opt. Phys.* **2020**, *53*, 155401. [CrossRef]
20. Jin, M.; Liu, J.; Xu, W.; Deng, D.; Han, L. Enhanced Goos-Hänchen Shift of SPR Sensor with TMDCs and Doped PANI/Chitosan Composites for Heavy Metal Ions Detection in Aquatic Environment. *Plasmonics* **2023**, *18*, 1129–1141. [CrossRef]
21. Lang, Y.; Liu, Q.; Wang, Q.; Zhou, X.; Jia, G. Wavelength-dependent Goos-Hänchen shifts observed in one-dimensional photonic crystal films with different structures. *Phys. Lett. A* **2022**, *449*, 128348. [CrossRef]
22. Olaya, C.M.; Hayazawa, N.; Hermosa, N.; Tanaka, T. Angular Goos-Hänchen Shift Sensor Using a Gold Film Enhanced by Surface Plasmon Resonance. *J. Phys. Chem. A* **2021**, *125*, 451–458. [CrossRef] [PubMed]
23. Sun, D. A proposal for digital electro-optic switches with free-carrier dispersion effect and Goos-Hanchen shift in silicon-on-insulator waveguide corner mirror. *J. Appl. Phys.* **2013**, *114*, 4502. [CrossRef]
24. Farmani, A.; Mir, A.; Sharifpour, Z. Broadly tunable and bidirectional terahertz graphene plasmonic switch based on enhanced Goos-Hänchen effect. *Appl. Surf. Sci.* **2018**, *453*, 358–364. [CrossRef]
25. Tsakmakidis, K.L.; Boardman, A.D.; Hess, O. 'Trapped rainbow' storage of light in metamaterials. *Nature* **2007**, *450*, 397–401. [CrossRef]
26. Liu, J.-Y.; Huang, T.-J.; Yin, L.-Z.; Han, F.-Y.; Liu, P.-K. High Sensitivity Terahertz Biosensor Based on Goos-Hnchen Effect in Graphene. *IEEE Photonics J.* **2020**, *12*, 1–6.
27. Palermo, G.; Sreekanth, K.V.; Maccaferri, N.; Lio, G.E.; Nicoletta, G.; De Angelis, F.; Hinczewski, M.; Strangi, G. Hyperbolic dispersion metasurfaces for molecularbiosensing. *Nanophotonics* **2021**, *10*, 295–314. [CrossRef]
28. Yuan, Y.; Peng, X.; Weng, X.; He, J.; Liao, C.; Wang, Y.; Liu, L.; Zeng, S.; Song, J.; Qu, J. Two-dimensional nanomaterials as enhanced surface plasmon resonance sensing platforms: Design perspectives and illustrative applications. *Biosens. Bioelectron.* **2023**, *241*, 115672. [CrossRef]
29. Sang, W.; Huang, S.; Chen, J.; Dai, X.; Liu, H.; Zeng, Y.; Zhang, T.; Wang, X.; Qu, J.; Ho, H.P.; et al. Wavelength sequential selection technique for high-throughput multi-channel phase interrogation surface plasmon resonance imaging sensing. *Talanta* **2023**, *258*, 124405. [CrossRef]

30. Zhang, C.; Hong, Y.; Li, Z.; Da, H. Giant and controllable Goos-Hnchen shift of monolayer graphene strips enabled by a multilayer dielectric grating structure. *Appl. Opt.* **2022**, *61*, 844–850. [CrossRef]
31. Benam, E.R.; Sahrai, M.; Bonab, J.P. High sensitive label-free optical sensor based on Goos-Hänchen effect by the single chirped laser pulse. *Sci. Rep.* **2020**, *10*, 17176. [CrossRef] [PubMed]
32. Ebert, M.; Weinbruch, S.; Hoffmann, P.; Ortner, H.M. The chemical composition and complex refractive index of rural and urban influenced aerosols determined by individual particle analysis. *Atmos. Environ.* **2004**, *38*, 6531–6545. [CrossRef]
33. Chew, W.C.; Weedon, W.H. A 3D perfectly matched medium from modified maxwell's equations with stretched coordinates. Microw. *Opt. Technol. Lett.* **1994**, *7*, 599–604. [CrossRef]
34. Chuang, S.L. Lateral shift of an optical beam due to leaky surface-plasmon excitations. *J. Opt. Soc. Am. A Opt. Image Sci.* **1986**, *3*, 593–599. [CrossRef]
35. Aminabhavi, T.M. Use of mixing rules in the analysis of data for binary liquid mixtures. *J. Chem. Eng. Data* **1984**, *29*, 54–55. [CrossRef]

photonics

MDPI

Article

Broadband Enhancement in the Spectral Response of Photovoltaic Modules with Flower-like Silver Particles

Yan Wang [1], Feng Zhang [2], Xinmin Fan [1,*], Yabin Lu [2], Chunyan Wang [1], Xiaodong Huang [1] and Lujun Zhang [1,*]

[1] Department of Physics and Optoelectronic Engineering, Weifang University, Weifang 261061, China; wfwy90@wfu.edu.cn (Y.W.); wangcy@bnu.edu.cn (C.W.); xdhuang@wfu.edu.cn (X.H.)

[2] Institute of Modern Optics, College of Electronic Information and Optical Engineering, Nankai University, Tianjin 300350, China; z.f.26@nku.edu.cn (F.Z.); luyabin@nku.edu.cn (Y.L.)

* Correspondence: xinminfan@wfu.edu.cn (X.F.); zhang_lujun1985@wfu.edu.cn (L.Z.)

Abstract: Recent research has indicated that metal nanoparticles, known for their unique optical properties, can enhance the spectral response of photovoltaic modules. Since most nanoparticles demonstrate enhancement effects within a specific wavelength range, broadening the spectral response of photoelectric devices is critical for their application in imaging, energy harvesting, and optical communication. In this study, we applied flower-like silver particles to achieve this broadband enhancement. The optical absorption of photovoltaic modules, featuring an amorphous Si p-i-n structure, was improved across a broad wavelength range of 400~2000 nm by integrating these flower-like silver particles, resulting in an approximately tenfold increase in peak spectral responsivity. The theoretical investigation further elaborates that the enhancement originates from the near-field effects of silver particles due to the interaction of different parts of the flower-like silver particles. Through these studies, we demonstrate that utilizing the flower-like silver particles with roughness surface can achieve the spectral response of the photoelectric device enhanced in broadband range, which can improve the utilization efficiency of optical energy for the applications of sensing, imaging, optical communication, and energy harvesting.

Keywords: broadband enhancement; photovoltaic; flower-like silver particles

1. Introduction

Due to their low cost and ease of fabrication, semiconductor photovoltaic modules have found widespread applications in both military and civilian industries [1–3]. A photodetector possessing a broadband spectral response holds significant implications for sensing, imaging, optical communication, and energy harvesting [4–8]. Therefore, developing a photodetector with a broadband spectral response has become a primary research area. Significant efforts have been directed towards improving the spectral response by modifying the material and structure of photovoltaic modules [9–11]. Over the past decade, with the advancement of plasmonics, enhancing the performance of photodetectors through the exploitation of metal nanostructures has been considered viable [12–15]. For instance, Naomi J. Halas and colleagues achieved enhanced spectral response in a photodiode within the range of 1250~1650 nm by using a gold (Au) antenna [16]. A substantial increase in the photoconductivity of amorphous silicon was also reported through the use of silica-coated gold nanorods [17]. Notably, metal nanoparticles with sizes up to 100 nm exhibit the most pronounced surface-localized field distribution. As we all know, the surface plasmon resonance of a single metal particle is primarily determined by factors such as particle composition, size, geometric shape, and environmental dielectric function. Particularly, metal particles with sizes within 100 nm exhibit the strongest surface localized field distribution. Recent studies have indicated that textured or defective surfaces on metal nanoparticles induce changes in the far-field and near-field optical properties of the nanoparticles [18]. Furthermore, when there's a coarse texture on the macroscopic metal surface or thin film, boundary conditions can be disrupted, enabling direct excitation of surface plasma waves. Our previous work systematically explored the fabrication and optical properties of flower-like silver nanoparticles. In our findings, we observed that these nanoparticles, characterized by their rough morphological features, exhibited broadband plasmon resonance peaks spanning across the visible and near-infrared regions of the spectra. Moreover, due to these specific characteristics of the flower-like silver nanoparticles, we achieved enhanced Surface-Enhanced Raman Scattering (SERS) intensity [19]. These results inspired us to design a nanoparticle-coupled semiconductor with the objective of achieving enhanced response over a broad spectral range. By providing this brief yet comprehensive summary of our previous work, we aim to give reviewers a clearer understanding of the basis and the objectives of our current study.

In this paper, we successfully fabricated an amorphous silicon (a-Si) photovoltaic module embedded with flower-like silver particles. We measured the transmittance, reflectance, and absorption characteristics of the spectra, with a particular focus on the spectral responses of the photovoltaic modules both with and without silver particles. Our findings show that, compared to modules without silver particles, the ones with silver particles demonstrate enhanced absorption and response across a broadband spectral range, extending from the visible to the near-infrared region. Furthermore, we examined the near-field optical properties of the flower-like silver particles with an average diameter of 500 nm. The results revealed that the interaction induced by the rough surface of the different components of the silver particles provides the flower-like silver particle with a unique field enhancement capability in the broadband range. This feature plays a pivotal role in the broadband enhancement of spectral response for the photovoltaic structure.

2. Materials and Methods

2.1. Preparation of Flower-like Silver Particles

Flower-like silver particles were synthesized using the standard chemical reduction method of silver nitrate, as reported in the existing literature [20,21]. All starting materials were of reagent grade and were used as received unless specified otherwise.

We added 0.6 mL of 1 M $AgNO_3$ and 6 mL of 100 mM polyvinylpyrrolidone aqueous solutions to 30 mL of pure water. The mixture was stirred continuously at room temperature, after which 0.6 mL of 1 M ascorbic acid was rapidly introduced. The stirring continued until no further color change was observed. After about eight minutes, the mixed solution turns dark gray in color and the reaction is complete. Following a centrifugation at 4000 rpm for 15 min, the flower-like silver particles were obtained. Additionally, since the silver particles obtained from the reaction inevitably retain some solvent residues, we introduce five times or more pure water into the silver particles. After centrifugation, we remove the supernatant using a pipette. Repeating this process three times to yield almost pure silver particles.

2.2. Preparation of a-Si p-i-n Photovoltaic Structure

The a-Si p-i-n photovoltaic structure was fabricated using plasma-enhanced chemical vapor deposition (RF-PECVD) technology. Firstly, ITO glass with a transparent conductive film was used as the substrate material and was cleaned using a high-temperature heating method. The ITO conductive glass was placed into the RF-PECVD device, the temperature was set to 160–180 °C, and it was maintained at a constant temperature for 2 h.

Then, the substrate was placed in three different reaction chambers and N, I, and P thin films were deposited in sequence. The reaction gas in the N chamber was silane, phosphane, and hydrogen gas, with a reaction power of 28 W and a film thickness of about 50 nm. The reaction gas in the I chamber was silane, germanane, and hydrogen gas, with a reaction power of 8 W and a film thickness of about 400 nm. The reaction gas in the P chamber was silane, borane, methane, and hydrogen gas, with a reaction power of 30 W and a film thickness of about 50 nm.

Finally, after the deposition of the three layers of thin film has been completed, the temperature drops to room temperature, and the substrate was removed from the RF-PECVD system, placed in the coating machine, and coated with metal electrodes.

2.3. Deposition of Flower-like Silver Particles

By using the self-assembly method, flower-like silver particles were deposited onto the photovoltaic structure. Firstly, the substrate (photovoltaic structure) was processed and ultrasonically cleaned in detergent for about 10 min. Then, it was ultrasonically cleaned in five times or more pure water for 10 min, and this step was repeated three times. Afterward, the substrate was cleaned in an ethanol solution, ultrasonically, for 10 min, followed by ultrasonic cleaning in a large amount of pure water for 10 min, this being repeated three times. After these steps, remove the substrate and dry it with nitrogen gas.

Next, attach a polymer layer. The polymer used here is 3-Aminopropyl trimethoxysilane. Mix it with pure water to obtain a polymer solution with a volume fraction of 1%. Soak the processed substrate in the solution for a period of time, remove it, and clean it with a methanol solution. Then, clean it with a large amount of ultrapure water and dry it with nitrogen gas.

Finally, proceed to the sedimentation process. Mix the prepared silver nanoparticles with five times or more pure water and stir thoroughly. Place the substrate in a sealed glove box filled with nitrogen gas and immerse it in the prepared flower-like silver particle sol solution for at least 2 h. Then we used a pipette to draw off the water on the surface. Afterward, take it out and gently blow-dry it with nitrogen gas.

2.4. Optical Modeling

Building upon our previous work, the flower-like silver particles consist of two components. One is the core particle, approximately 400 nm in size, and the other consists of surface protrusions, each around 100 nm in size. This design ensures that the total size of the particle is 500 nm [19], as illustrated in Figure 1.

Figure 1. Schematic diagrams of flower-like silver particles.

The surface local fields of the silver particles were calculated using the three-dimensional finite difference time domain (FDTD) method. The dielectric data for silver was adopted from Palik [19,22]. A perfectly matched layer (PML) was employed as the boundary condition. The excitation light was set to be incident in the positive z-direction and polarized along the x-axis.

2.5. Instrumentation

The photovoltaic structure was prepared using a PlasmaPro 800Plus RF-PECVD device from Oxford Instruments. Scanning electron microscopy (SEM) images of the specimens were acquired with a JEOL JSM-6700f scanning electron microscope at 3.0 kV. The transmittance and reflectance spectra were recorded using a Cary500 UV-visible-infrared absorption spectrometer. The photocurrent was assessed with a Keithley 6430 digital source meter.

3. Results and Discussion

The flower-like silver particles were deposited onto the surface of an a-Si p-i-n structure, which was immobilized on ITO glass, and then left to dry naturally in a nitrogen environment. As shown in Figure 2a, the thickness of the a-Si p-i-n structure is measured to be 500 nm. From the SEM image shown in Figure 2b, it can be observed that the silver particles, with an average diameter of ~500 nm, are distributed randomly on the surface of the a-Si p-i-n structure.

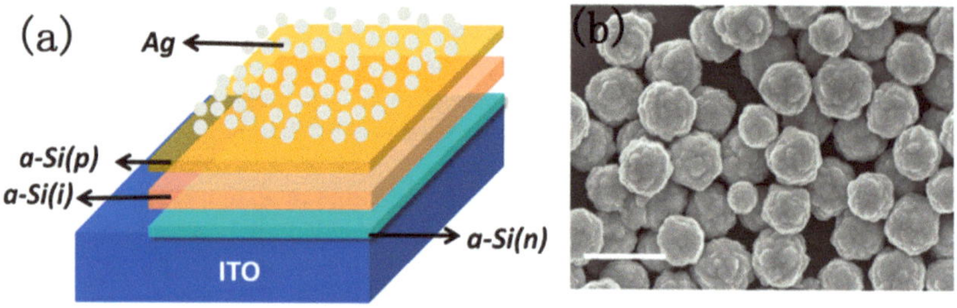

Figure 2. (**a**) Schematic of the a-Si p-i-n photovoltaic structure with silver particles on the surface. (**b**) The SEM image of the flower-like silver particles on the surface (the scale bar = 1 μm).

The transmittance (Figure 3a) and reflectance (Figure 3b) of the a-Si p-i-n photovoltaic structure, both with and without silver particles, were respectively measured using the diffuse reflection method. Absorptance, defined as Abs (%) = 1 − R (%) − T (%), is depicted in Figure 3c. For the sample adorned with silver particles, a decrease in transmittance is noted across the wavelength range of 450~2000 nm, particularly between 450 nm and 1600 nm. As for reflectance, it decreased across the full wavelength range of 200~2000 nm when silver particles were present on the photovoltaic structure. Hence, absorptance was enhanced across the entire wavelength spectrum, from 200 nm to 2000 nm. A comparison of the spectral curves of the a-Si p-i-n structure with and without silver particles reveals no change in the shape of absorption. This phenomenon is attributed to the flower-like silver particles possessing broadband plasmon resonance peaks in the visible and near-infrared regions of the spectra, thereby enhancing absorption across the entire wavelength range for the photovoltaic module. A comparison of the spectra of the semiconductor with and without flower-like silver particles demonstrates that the spectral response of the a-Si p-i-n structure can be enhanced through the incorporation of flower-like silver particles.

Then, the photocurrent in the wavelength range of 400~800 nm of the a-Si p-i-n photovoltaic structure with and without silver particles was further measured, respectively. The responsivity was calculated by $R = I_L/P_{inc}$, where R is defined as the responsivity, I_L is the photocurrent, and P_{inc} is the power of the incident light. We plotted the responsivity R as a function of wavelength, as shown in Figure 4. The results show that the structure without silver particles has a spectral response at the wavelength range of 400 nm to 800 nm, which is consistent with the results in the previous report [23]. Compared to the structure without silver particles, the spectral response of the composite structure has been obviously enhanced. Especially, at the wavelength of 650 nm, i.e., at the peak of the responsivity, the responsivity was enhanced by about 10 times. At the same time, the spectral response shape was consistent with that of the structure without silver particles.

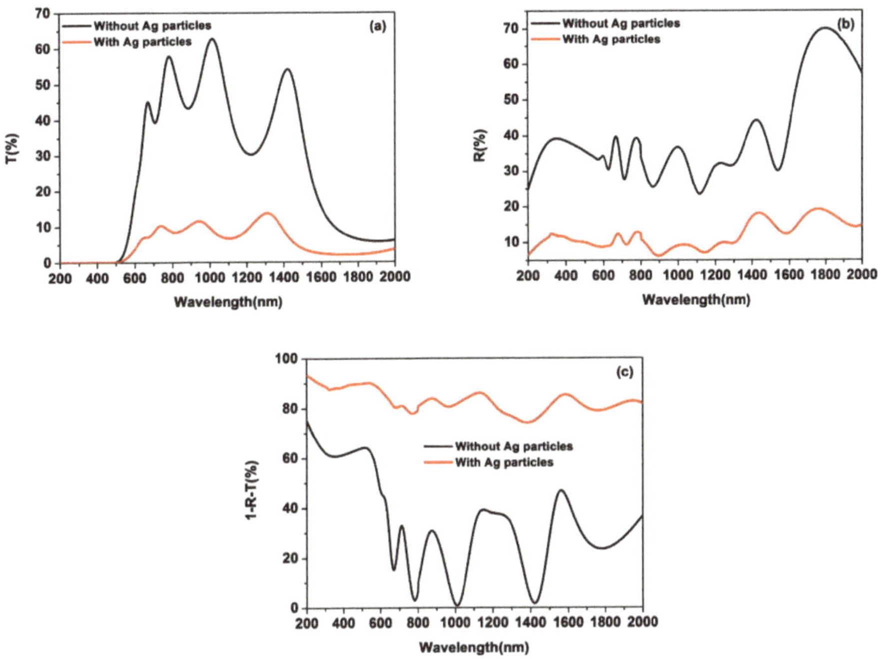

Figure 3. The spectra measured for (**a**) the transmittance, (**b**) the reflectance, and (**c**) the absorptance, respectively.

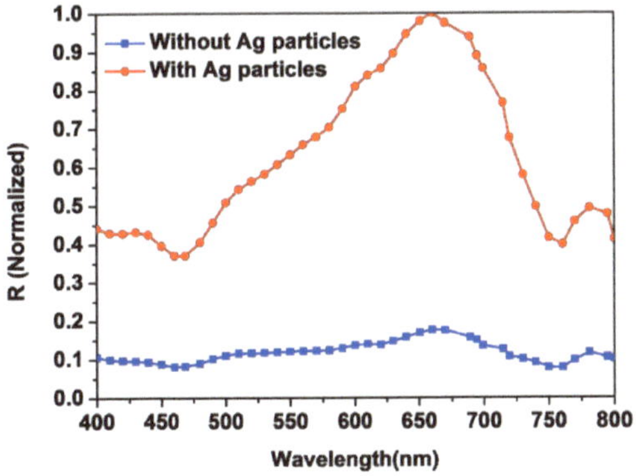

Figure 4. The responsivity of the a-Si p-i-n structure with and without silver particles, respectively.

For a semiconductor, the responsivity $R_{\lambda,I}$ is proportional to the intensity of the radiation field and can be expressed as

$$R_{\lambda,I} \propto C\left|\vec{E}\right|^2 \quad (1)$$

where the parameter C represents a constant. Consequently, an increase in the intensity of the radiation field leads to an enhancement of the spectral response. In our photovoltaic structure with silver particles, the incident light initially interacts with the flower-like silver particles deposited on the surface of the a-Si p-i-n structure, stimulating the localized surface plasmon resonance of the silver particles and enhancing the field around them. This interaction consequently amplifies the intensity of the radiation field of the semiconductor.

The results shown in Figure 3 demonstrate that the flower-like silver particles, with an average diameter of 500 nm, can enhance the spectral response across the full wavelength range of 400~800 nm. This suggests that the flower-like silver particles possess a distinctive field enhancement capability within the 400~800 nm wavelength range. Therefore, it is believed that the surface roughness structure of the flower-like silver particle plays a pivotal role in contributing to field enhancement.

We have previously published papers on enhancing the fluorescence signal of polycyclic aromatic hydrocarbons in diesel using silver nanoparticles [24]. The experimental results and theoretical analysis indicate that due to the presence of silver nanoparticles, the absorption of excited light by polycyclic aromatic hydrocarbons is enhanced, thereby increasing the fluorescence signal by 4.6 times. Combining the experimental results in Figures 3 and 4, it can be concluded that silver nanoparticles enhance the absorption of photovoltaic devices.

To further investigate the influence of the surface protrusions, the flower-like silver particle was considered to comprise of two components: a large core particle of ~400 nm in size, and smaller surface particles of ~100 nm, as shown in Figure 1. The plasmon resonance mode for the surface particles maintains the form of a dipole consistently across the 400 nm to 800 nm wavelength range. However, for the core particle, the plasmon resonance mode starts as a multipole at short wavelengths, transitioning to a dipole at longer wavelengths [25].

The FDTD method was subsequently utilized to simulate the changes in local field intensity and distribution of flower-like silver particles and the central large sphere at wavelengths of long wave (650 nm) and short wave (410 nm), as depicted in Figure 5. Under the excitation of light at a wavelength of 410 nm, a localized field distribution of multipole plasma resonance appears on the surface of the central sphere, as illustrated in Figure 5a. This observation aligns with the Mie theory calculation results. As the excitation wavelength shifts to 650 nm, the excitation wavelength falls within the range of dipole plasmon resonance scattering peaks of silver particles. Consequently, the surface local field exhibits a dipole plasmon resonance distribution, as presented in Figure 5b. According to the foundational theory of metal surface plasmon resonance, when the size of metal nanoparticles exceeds 100 nm, an increase in particle size will diminish their surface enhancement effect. Thus, for a central sphere with a diameter of 400 nm, the local surface field is significantly reduced due to particle size limitations. For flower-like silver particles measuring 500 nm in size, the local field distribution is showcased in Figure 5c,d. In contrast to the central sphere, the local field intensity of the flower-like silver particles has markedly improved. At excitation wavelengths of 410 nm and 650 nm, the maximum local field enhancements are 35 times and 6.3 times, respectively. Concurrently, the local field distribution on the surface of flower-like silver particles exhibits notable changes compared to the distribution on the central sphere. With an excitation wavelength of 410 nm, the local field of flower-like silver particles is primarily distributed between the grooves of the surface structure. At 650 nm, the most pronounced localized field emerges in the peripheral region of the flower-like silver particles, oriented parallel to the polarization direction of the incident light. The local field distribution of silver particles is predominantly dictated by their plasmon resonance mode. Comparing the local field distribution of flower-like silver particles to that of the central sphere reveals that while the distribution patterns of the two particle types are similar, the presence of edge balls substantially amplifies the local field strength surrounding the flower-like silver particles.

In summary, the plasma resonance mode of submicron silver particles is similar to that of the central sphere, with multipole resonance in the short wave and dipole resonance in the long wave. The vibration of the internal electrons is also mainly determined by the central large sphere. Therefore, we investigated the local field distribution of the edge sphere under the action of the central sphere at excitation wavelengths of 410 nm and 650 nm, respectively. Figure 6 depicts the local field of three neighboring surface particles stimulated by the local field of the core particle. Under the 410 nm radiation, the strongest field for the surface particles occurs in the gap of the surface roughness structure, as depicted in Figure 6a. Under 650 nm wavelength radiation, the electric field of surface particles is concentrated on the outer area of the surface, as presented in Figure 6b. These simulation results disclose that the effects of the core particle on the surface particles differ in the short and long wavelength ranges, and the "hot spot" of the surface particles also varies accordingly. Furthermore, the interaction between surface particles was also examined. It is found that the electron distribution of the surface particles is determined by the core particle, considered as a whole.

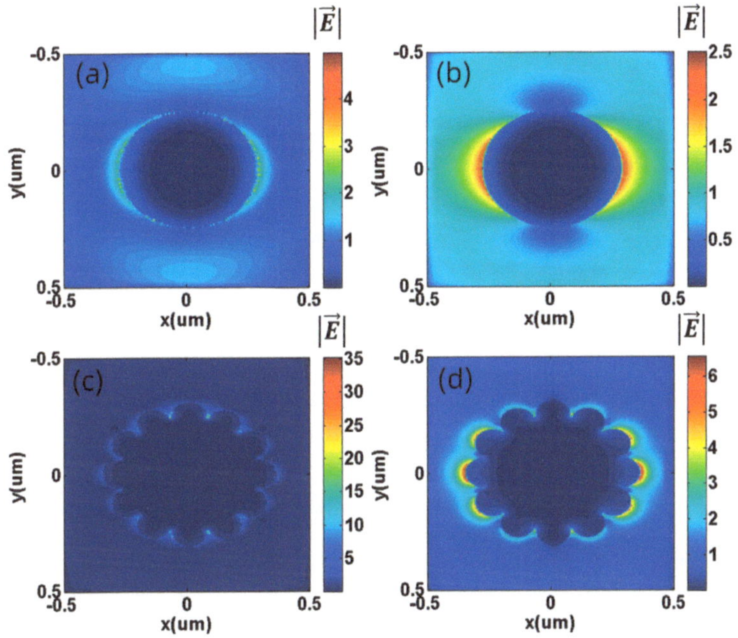

Figure 5. The electric field distribution of different silver nanoparticle models (**a**,**b**) corresponds to the 400 nm spherical silver nanoparticles model excited at 410 nm and 650 nm, respectively; (**c**,**d**) correspond to the flower-like silver nanoparticles model excited at 410 nm and 650 nm, respectively.

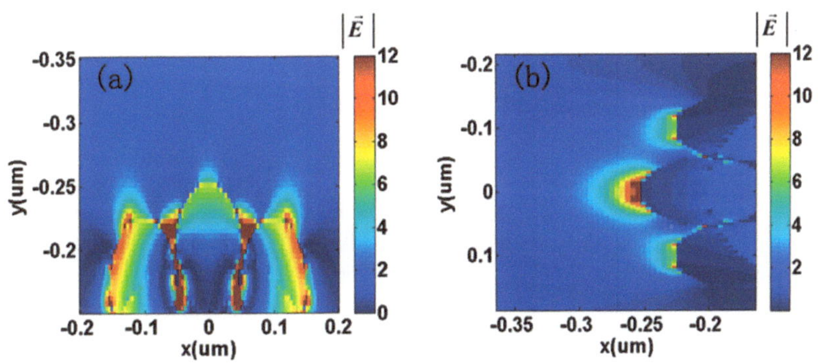

Figure 6. The distribution of the electric field around three neighboring surface particles excited by the local field of the core particle. (**a**) at the wavelength of 410 nm, and (**b**) at the wavelength of 650 nm.

For a single surface particle, the resonance model is solely a dipole in the wavelength range of 400~800 nm and can therefore be considered as a dipole. Under the radiation at a wavelength of 410 nm, the resonance model of the core particle is multipole. The internal electron oscillations are not just parallel but also perpendicular to the polarization of the incident light. As a result, for neighboring particles in the "Shoulder to Shoulder" pattern (the axis of which is parallel to the polarization of the incident light) of the surface particles, the interaction between electrons can occur [26,27], as illustrated in Figure 7a. The strongest electric field occurs in the gap between the two particles which are in the "Shoulder to Shoulder" pattern.

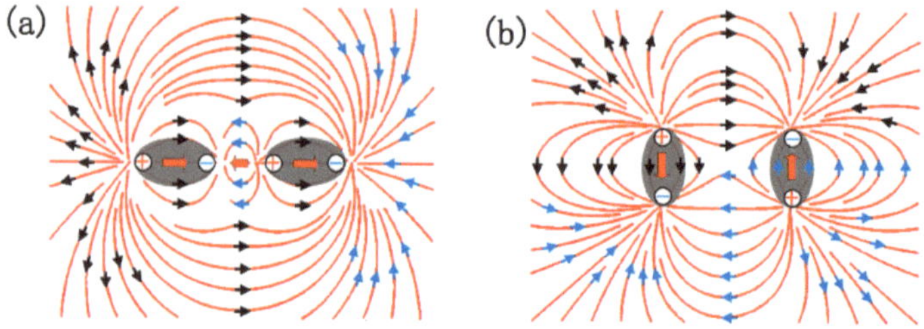

Figure 7. A sketch of the electric field distribution of two neighboring surface particles. (**a**) "Shoulder to Shoulder" pattern, under the action of the core particle at 410 nm. (**b**) "Head-to-Head" pattern, under the action of the core particle at 650 nm wavelength.

Under radiation at a wavelength of 650 nm, the resonance model of the core particle is a dipole, and collective electron oscillations are parallel to the direction of the incident light. The internal electrons concentrate on both sides of the core particle. Therefore, the interaction of surface particles in the "Shoulder to Shoulder" pattern is suppressed. Conversely, for neighboring particles in the "Head-to-Head" pattern (the axis of which is perpendicular to the polarization of the incident light) among the surface particles, the local field is enhanced under the action of the core particle, as depicted in Figure 7b. In this case, the electric field is concentrated in the outer area.

The local field of the surface particles reciprocally impacts the local field of the core particle. The electric field distribution of the core particle, under the influence of the local field of the surface particles, is calculated using the FDTD method, and the results are presented in Figure 8. Figure 8a displays the electric field distribution of the core particle when the surface particles are in the "Shoulder to Shoulder" configuration, with an excitation wavelength of 410 nm. Influenced by the surface particles, the local field of the core particle is enhanced, especially in the gap between the particles. Figure 8b demonstrates the electric field distribution of the core particle when the surface particles are in a "Head-to-Head" arrangement, with an excitation wavelength of 650 nm. The most intense field of the core particle is primarily located in the outer area of the particle, due to the presence of the surface particles.

In comparison to the local field distribution of smooth spheres in the incident light, the local field of the flower-like silver particle is repositioned due to the surface particles and is significantly amplified in both the short and long wavelength ranges. This result aligns with our previous findings on flower-like silver particles [19,25].

The interaction between the surface particles and the core particles allows the flower-like silver particles to achieve a near-field enhancement across a broad spectrum. The field enhancement is defined as $|E|/|E_0|$ and is plotted as a function of wavelength in Figure 8. Here, $|E_0|$ is the incident field and $|E|$ is the maximum local field of the flower-like silver particle.

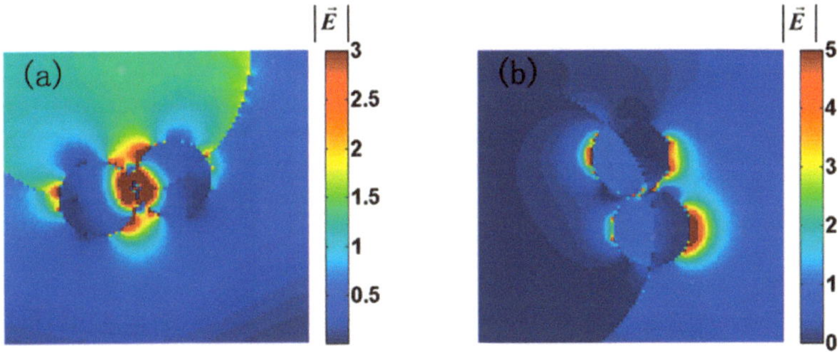

Figure 8. The electric field distribution of core particles under the effect of the local field of surface particles. (**a**) at the wavelength of 410 nm, (**b**) at the wavelength of 650 nm.

Figure 9 reveals that the field enhancement of the flower-like particle persists across a broad wavelength range of 400~800 nm, with the field enhancement being more significant in the shorter wavelength range. However, as the local field distribution is concentrated in the gap of the surface particles in the short wavelength range, the enhancement ratio of the spectral responsivity is conversely lower than in the long wavelength range.

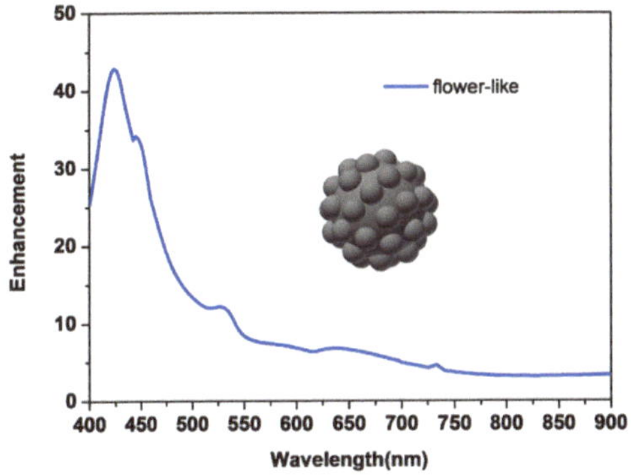

Figure 9. The local field enhancement $|E|/|E_0|$ of the flower-like silver particles to the wavelength.

After the incident light acts on the silver particles, due to the LSPR effect of the silver particles, the radiation field acting on the semiconductor is enhanced. After the semiconductor absorbs the photons, the internal electronic transition process can be explained by the perturbation theory in quantum mechanics [28,29]. According to Fermi's Golden Rule, the transition probability w_{if} of the semiconductor can be expressed as:

$$w_{if} = C\left|\vec{E}\right|^2 |M_{if}|^2 \quad (2)$$

where C is a constant, M_{if} represents the transition matrix element, $\vec{E}$ is the radiation field acting on the semiconductor. The relationship between the semiconductor's absorption coefficient α and the transition probability w_{if} can be expressed as:

$$\alpha \propto \sum w_{if} n_i n_f \quad (3)$$

where n_i and n_f represent the density of states of the semiconductor in the initial and final state, respectively. According to Equations (2) and (3), for amorphous silicon photovoltaic modules containing silver particles, the action of the silver particles enhances the radiation field acting on the semiconductor, thereby increasing the probability of internal electronic transitions in the semiconductor and raising the absorption coefficient. Generally speaking, for the same semiconductor, its absorption coefficient is a fixed value, but the introduction of silver particles can be expressed as an increase in its absorption coefficient. Thus, when the light of the same intensity acts on amorphous silicon photovoltaic modules with and without silver particles, the photovoltaic modules containing silver particles have a larger absorption coefficient, therefore their light absorption is stronger, resulting in a larger photocurrent.

4. Conclusions

In summary, an a-Si p-i-n photovoltaic structure with flower-like silver particles deposited on the surface was fabricated. The transmittance, reflectance, and absorptance spectra measurements revealed that the absorption of this photovoltaic structure was enhanced in a broad wavelength range from 200 nm to 2000 nm by adding flower-like silver particles on the surface. Furthermore, the spectral response of the a-Si p-i-n structure with and without silver particles, respectively, was measured at the wavelength range of 400~800 nm, which revealed that the maximum enhancement ratio of the responsivity can reach about 10, and the spectral response shape of the composite structure was consistent with the structure without silver particles. The FDTD method was employed to investigate the mechanism. In our analysis, the sample was divided into two parts including the core particle and surface particles. We found that the interaction between the core particle and surface particles leads to the unique optical characteristics of the flower-like particles, which improved the performance of the a-Si p-i-n photovoltaic structure in a broadband range. Through these studies, we demonstrate that utilizing the subwavelength silver particles with roughness surface can achieve the spectral response of the photoelectric device enhanced in broadband range, which can improve the utilization efficiency of optical energy for the applications of sensing, imaging, optical communication, and energy harvesting. The theoretical analysis provides improved insight into the coupling of physical properties in photovoltaic modules incorporating metal particles.

Author Contributions: Y.W.: Methodology, Formal analysis, Date curation. F.Z.: Investigation, Software. X.F.: Funding acquisition, Writing—review & editing. Y.L.: Methodology, Data curation. C.W.: Investigation. X.H.: Software, Supervision. And L.Z.: Software, Funding acquisition. All authors have read and agreed to the published version of the manuscript.

Funding: This research was funded by the National Natural Science Foundation of China under Grant 61805178, and in part by the Natural Science Foundation of Shandong Provincial under Grant ZR2019MD011 and ZR2020QA072.

Institutional Review Board Statement: Not applicable.

Informed Consent Statement: Not applicable.

Data Availability Statement: Not applicable.

Conflicts of Interest: The authors declare no conflict of interest.

References

1. Wong, J.; Omelchenko, S.T.; Atwater, H.A. Impact of Semiconductor Band Tails and Band Filling on Photovoltaic Efficiency Limits. *ACS Energy Lett.* **2021**, *6*, 52–57. [CrossRef]
2. Harikesh, P.C.; Surendran, A.; Ghosh, B.; John, R.A.; Moorthy, A.; Yantara, N.; Salim, T.; Thirumal, K.; Leong, W.L.; Mhaisalkar, S.; et al. Cubic NaSbS2 as an ionic–electronic coupled semiconductor for switchable photovoltaic and neuromorphic device applications. *Adv. Mater.* **2020**, *32*, 1906976. [CrossRef] [PubMed]
3. Liu, X.; Zhou, S. Progress on photovoltaic AlGaN photodiodes for solar-blind ultraviolet photodetection. *Chin. Opt. Lett.* **2022**, *20*, 112501. [CrossRef]
4. Ren, Y.; Van, V. Enhanced Small-Signal Responsivity in Silicon Microring Photodetector Based on Two-Photon Absorption. *IEEE J. Sel. Top. Quantum Electron.* **2020**, *26*, 1–8. [CrossRef]

5. Yang, X.; Qu, L.; Gao, F.; Hu, Y.; Yu, H.; Wang, Y.; Cui, M.; Zhang, Y.; Fu, Z.; Huang, Y.; et al. High-Performance Broadband Photoelectrochemical Photodetectors Based on Ultrathin Bi2O2S Nanosheets. *ACS Appl. Mater. Interfaces* **2022**, *14*, 7175–7183. [CrossRef] [PubMed]
6. Cao, F.; Meng, L.; Wang, M.; Tian, W.; Li, L. Gradient energy band driven high-performance self-powered perovskite/CdS photodetector. *Adv. Mater.* **2019**, *31*, 1806725. [CrossRef] [PubMed]
7. Liu, S.B.; Chang, S.J.; Chang, S.P.; Chen, C.H. An Amorphous (Al0.12Ga0.88)$_2$O$_3$ Deep Ultraviolet Photodetector. *IEEE Photonics J.* **2020**, *12*, 1–8. [CrossRef]
8. Dong, Q.; Wang, F.; Hu, X.; Lu, Y.; Zhao, D.; Zhang, M. High-performance broadband photodetector based on PdSe2/black phosphorus heterodiode. *Appl. Phys. Lett.* **2022**, *120*, 23. [CrossRef]
9. Zhong, Z.; Xu, Z.; Sheng, T.; Yao, J.; Xing, W.; Wang, Y. Unusual air filters with ultrahigh efficiency and antibacterial functionality enabled by ZnO nanorods. *ACS Appl. Mater. Interfaces* **2015**, *7*, 21538–21544. [CrossRef]
10. Kim, Y.J.; Yoo, Y.J.; Yoo, D.E.; Lee, D.W.; Kim, M.S.; Jang, H.J.; Kim, Y.-C.; Jang, J.-H.; Kang, I.S.; Song, Y.M. Enhanced light harvesting in photovoltaic devices using an edge-located one-dimensional grating polydimethylsiloxane membrane. *ACS Appl. Mater. Interfaces* **2019**, *11*, 36020–36026. [CrossRef]
11. Sun, X.; Su, Y.; Huang, Y.; Tan, J.; Yi, J.; Hu, T.; Zhu, L. Photovoltaic Modules Monitoring Based on WSN With Improved Time Synchronization. *IEEE Access* **2019**, *7*, 132406–132412. [CrossRef]
12. Zhong, J.H.; Vogelsang, J.; Yi, J.M.; Wang, D.; Wittenbecher, L.; Mikaelsson, S.; Korte, A.; Chimeh, A.; Arnold, C.L.; Schaaf, P. Nonlinear plasmon-exciton coupling enhances sum-frequency generation from a hybrid metal/semiconductor nanostructure. *Nat. Commun.* **2020**, *11*, 1464. [CrossRef] [PubMed]
13. Abbasi, M.; Evans, C.I.; Chen, L.; Natelson, D. Single Metal Photodetectors Using Plasmonically-Active Asymmetric Gold Nanostructures. *ACS Nano* **2020**, *14*, 17535–17542. [CrossRef] [PubMed]
14. Wang, K.; Niu, L.; Tao, L.; Zhang, Y.; Zhou, X. Fabrication of tio_2 microspheres with continuously distributed sizes from nanometer to micronmeter: The increasing light scattering ability and the enhanced photovoltaic performance. *Sol. Energy* **2021**, *11*, 230.
15. Mondal, A.; Yadav, M.K.; Shringi, S.; Bag, A. Extremely low dark current and detection range extension of Ga$_2$O$_3$ UV photodetector using Sn alloyed nanostructures. *Nanotechnology* **2020**, *31*, 294002. [CrossRef] [PubMed]
16. Knight, M.W.; Sobhani, H.; Nordlander, P.; Halas, N.J. Photodetection with active optical antennas. *Science* **2011**, *332*, 702–704. [CrossRef]
17. Chang, C.-S.; Rothberg, L.J. Plasmon-enhanced photoconductivity in amorphous silicon thin films by use of thermally stable silica-coated gold nanorods. *Chem. Mater.* **2015**, *27*, 3211–3215. [CrossRef]
18. Wang, H.; Goodrich, G.P.; Tam, F.; Oubre, C.; Nordlander, P.; Halas, N.J. Controlled Texturing Modifies the Surface Topography and Plasmonic Properties of Au Nanoshells. *J. Phys. Chem. B* **2005**, *109*, 11083–11087. [CrossRef]
19. Zhang, F.; Chen, P.; Zhang, L.; Mao, S.-C.; Lin, L.; Tang, Y.-G.; Cui, J.-C.; Qi, X.-D.; Yang, J.-H.; Ma, Y.-F. Enhancement of Raman scattering by field superposition of rough submicrometer silver particles. *Appl. Phys. Lett.* **2012**, *100*, 173103. [CrossRef]
20. Liang, H.; Li, Z.; Wang, W.; Wu, Y.; Xu, H. Highly surface-roughened "flower-like" silver nanoparticles for extremely sensitive substrates of surface-enhanced Raman scattering. *Adv. Mater.* **2009**, *21*, 4614–4618. [CrossRef]
21. Kneipp, K.; Kneipp, H.; Itzkan, I.; Dasari, R.; Feld, M.S. Surface-enhanced Raman scattering and biophysics. *J. Phys. Condens. Matter* **2002**, *14*, 597–624. [CrossRef]
22. Prucha, E.J.; Palik, E.D. *Handbook of Optical Constants of Solids*; Academic Press: Cambridge, MA, USA, 1998.
23. Carlson, D.E.; Wronski, C.R. Amorphous silicon solar cell. *Appl. Phys. Lett.* **1976**, *28*, 671–673. [CrossRef]
24. Zhang, F.; Zhang, L.; Mao, S.C.; Chen, P.; Cui, J.C.; Tang, Y.G.; Wang, K.; Lin, L.; Qi, X.D. Use of metal-enhanced fluorescence spectroscopy for detection of polycyclic aromatic hydrocarbons in diesel oil emulsions in artificial seawater. *Environ. Technol.* **2012**, *33*, 2071–2075. [CrossRef] [PubMed]
25. Zhang, F.; Chen, P.; Li, X.; Liu, J.T.; Lin, L.; Fan, Z.W. Further localization of optical field for flower-like silver particles under laser radiation. *Laser Phys. Lett.* **2013**, *10*, 045901. [CrossRef]
26. Tong, L.; Wei, H.; Zhang, S.; Li, Z.; Xu, H. Optical properties of single coupled plasmonic nanoparticles. *Phys. Chem. Chem. Phys.* **2013**, *15*, 4100–4109. [CrossRef]
27. Jain, P.K.; Huang, W.; El-Sayed, M.A. On the universal scaling behavior of the distance decay of plasmon coupling in metal nanoparticle pairs: A plasmon ruler equation. *Nano Lett.* **2007**, *7*, 2080–2088. [CrossRef]
28. Fox, M. *Quantum Processes in Semiconductors*, 5th ed.; University Press: Oxford, MS, USA, 2014; Volume 55, pp. 338–339.
29. Fan, R.; Liu, Z.; Jin, D.; Luo, T.; Li, N.; Li, S.; Wang, Y.; Xia, Y.; Lu, Z. High temporal waveform fidelity stimulated Brillouin scattering phase conjugate mirror using Novec-7500. *Opt. Express* **2023**, *31*, 1878–1887. [CrossRef]

MDPI AG
Grosspeteranlage 5
4052 Basel
Switzerland
Tel.: +41 61 683 77 34

Photonics Editorial Office
E-mail: photonics@mdpi.com
www.mdpi.com/journal/photonics

www.ingramcontent.com/pod-product-compliance
Lightning Source LLC
Chambersburg PA
CBHW041342210926
51850CB00013B/2955